Politics, Pitchforks and Pickle Jars

75 Years of Organized Farm Women in Alberta

Nanci Langford

[signature: Nanci Langford]

Detselig Enterprises Ltd.

Calgary, Alberta, Canada

Sponsored by Women of Unifarm

Politics, Pitchforks and Pickle Jars

Canadian Cataloguing in Publication Data

Langford, Nanci L., 1954-
  Politics, pitchforks and pickle jars

Includes index.
ISBN 1-55059-147-9

1. Women of Unifarm (Association)—History.   2. Women in agri-
culture—Alberta—History.   3. Women in politics—Alberta—History.
I. Title.
HQ1909.A4L36 1997   305.4'06'07123   C97-910329-0

Detselig Enterprises Ltd.
210-1220 Kensington Rd. N.W.
Calgary, Alberta T2N 3P5

Detselig Enterprises Ltd. appreciates the financial support for our 1997
publishing program, provided by Canadian Heritage and the Alberta
Foundation for the Arts, a beneficiary of the Lottery Fund of the Government of Alberta.

Printed in Canada
ISBN 1-55059-147-9
SAN 115-0324

Cover design by Dean Macdonald, based on contest-winner Jenny
Stirling's cover design.

Table of Contents

Preface

Many books have an opening statement "characters in this book are entirely fictional, and the facts herein are not factual or historical. No inference or conclusion to living persons or events may be drawn therefrom." We start our book, *Politics, Pitchforks and Pickle Jars*, from an entirely different perspective. The characters you will meet herein were living people – dedicated people – people deeply and sincerely committed to the betterment of rural life in this province of Alberta. Then, as transportation and communication systems expanded far beyond our wildest dreams, we realized we were part of a national, and then a global society, and we expanded our horizons to world betterment. So the events recorded in this book are indeed historical and hopefully as you peruse its pages you will find much of interest.

We commend the Women of Unifarm for undertaking to write a history book which goes back 75 years to include the early development of the farm women's movement in Alberta. Farm women were members of early farm organizations. They, along with the men, provided leadership in the United Farmers of Alberta (UFA) prior to 1915, when the United Farm Women of Alberta was formed as a separate women's section. The women of the Alberta Farmers' Union (AFU) provided leadership, working with the men without a separate women's organization. In 1949, when the UFA and the AFU amalgamated to form the Farmers' Union of Alberta, the women from both organizations brought excellent leadership into the newly organized Farm Women's Union of Alberta and carried that leadership into the Women of Unifarm.

We, Louise Johnston and Betty Pedersen, feel that we represent not only the many able women who preceded and followed us in leadership, but also the women of the locals, for our work began at our own farm gates. Ours was an organization that grew and changed to meet the needs and challenges of the changing rural society in which we lived. Changes are usually slow in coming, and we had many periods of worry and frustration and then our brief moments of glory when our efforts bore fruit.

We were members of many boards, committees and commissions far beyond the scope of our own organization, and from them we brought reports and information to our own board, whose duty it was to channel it back to our membership. We thought of ourselves as a pipeline, so that our members could present well-documented resolutions that might lead to community action or the legislative changes we desired. We tried to offer good leadership and organization, and sometimes, to be truthful, we wondered at our own temerity as we delved into matters of education, health and all the social aspects of rural living.

We are pleased to share our history with you, and we commend the Women of Unifarm History Book Steering Committee and the author, Nanci Langford, for their research work and their organization of the material into an interesting book. We appreciate the inclusion of original documents and the useful references to sources. So join us now as we travel the length and breadth of our province, and far beyond, to look at rural women, and salute their efforts as they strove to raise their children, feed their families, support their men, participate in cooperative living and work whole-heartedly for church and community.

Louise Johnston   *President of Farm Women's Union of Alberta 1963-1967*

Elizabeth (Betty) Pedersen   *President of Farm Women's Union of Alberta/ Women of Unifarm, 1969-1974*

Acknowledgements

This book began with the expressed wish of many Alberta farm women to have their story told. Margaret Blanchard, as outgoing President of Women of Unifarm in 1989, provided the push that started the project, supported by a motion made at a board meeting by Verna Kett. Janet Walter provided the leadership, picking up the task of seeing the book through from start to finish over a period of six years. Janet assembled a steering committee composed of Corinne Thompson, Joyce Templeton, Janet Walter, Louise Christiansen, Elizabeth Olsen, Norah Keating and the author Nanci Langford. Louise Christiansen and Elizabeth Olsen were later replaced on the steering committee by Verna Kett.

The financial support of the Women's Program, the Secretary of State, the Canadian Research Institute for the Advancement of Women, the United Farmers of Alberta and the Alberta Historical Resources Foundation is gratefully acknowledged. Women of Unifarm sponsored the cash prize for the cover contest, and Betty Pedersen sponsored the cash prize for the title contest.

The author was ably assisted by two research assistants, Sheila Dunphy and Kammi Rosentreter. Kammi devoted extra unpaid hours to the project and was an enthusiastic supporter of the book. Barbara Demers provided meticulous copy editing and excellent editorial comments. Corinne Thompson and Joyce Templeton selected and arranged for the photographs in the book. The outstanding cover design is based on the creation of Jenny Stirling of Westlock, and the book's title was the inspiration of Mabel Sorensen of Peace River.

This history would not be possible without the memories and personal collections of the women of the organization. For the donation of materials and for their personal interviews, we thank Rita Graumans; Elizabeth Kocsis; Katie Dyck; Eleanor Trotter; Irma Merkl; Sophie Rusk; Lois Bunney; Natalie Albereta; Marion Alberts; Donna Meador; Dorothy Houseman; Nettie Robinson; Doreen Sexty; Betty Seatter; Helen Gray; Mary Osadczuk; Mary Wright; Mabel Legrange; Edna Butler; Maisie Jacobson; Irene Wagstaff; Marguerite Mason; Janet Walter; Florence Trautman; Joyce Templeton; Helen Murray; Dorothy Kuehn; Verna Kett; Louise Christiansen; Emma Innocent; Naomi Findlay; Rita Cannard; Mabel Barker; Sonya Hudson; Verna Messner; Nettie Connelly; Margaret Ann Alexander; Corinne Thompson; Margaret Blanchard; Dorothy Ottewell; Miriam Galloway; Elizabeth Olsen; Ruth Wilson; Jacqueline

Galloway, Betty Pedersen; Louise Johnston; Hazel Andersen; Teresa Maykut; Olwen Murray; Mary Belanger; Lena Haywood; Evelyn Long; Jean Ross; Florence Sheard; Inga Marguardt; Lena Scraba; Alice Woychuk; Barbara Klymchuk; Maisy Platt; Jenny Stirling; Jean Mansfield; Isabel Russell; Alberta Keinholz; Eva Byvank; Florence Crawford; Florence Scissons; Blanche Hoar; Jean Rose: Betty Matjeka; Adelaide Dootson; Peggy Jackson; Jean Culler; Jean Buit; Margaret House; Greta Hallett; Doris Barker; Mildred Pollock; Dorothy Clayton; Joyce McElray; Edith Braithwaite; Colleen Casey Cyr.

A big thank you also to all the women who arranged interviews and who conducted interviews in various regions of the province, with special recognition to Margaret Allan, Judy Pimm and Janet Walter.

Many locals and individual women put together their own histories or scrapbooks that were extremely useful in compiling this history. We sincerely appreciate their efforts and their support. The Unifarm office staff were always helpful and pleasant whenever called upon for information.

The author would like to thank Sylvia McKinlay and Rick Vanden Ham, who provided moral and practical support throughout the years of research and writing. She also appreciated the warm welcome and wonderful refreshments she received as she toured the province to interview farm women and gather materials.

Finally, the history book committee would like to acknowledge the extraordinary contributions of organized farm women in all organizations to the quality of life in Alberta. To these women of foresight, determination and action, we all owe a debt of gratitude and recognition.

Introduction

This is an Alberta story of importance to all Albertans. Alberta was built on the prosperity created by the hard work and dedication of farmers and ranchers. Generations of farm men and women not only made a living from the land, but also made this land a good place to live. We are indebted to the farm men and women who, through their determination, their foresight and their courage, created communities and came together as communities to build a province.

This book records the role of organized farm women in the development of the province of Alberta. Through an organization named United Farm Women of Alberta (UFWA), renamed Farm Women's Union of Alberta in 1949, and, in 1970, Women of Unifarm, thousands of women have joined together over the years to try to create a society of progress and possibilities, for themselves, their neighbors and their children. Founded on the belief that agriculture is the foundation of economic and social prosperity and stability, and with the conviction that what benefits farmers benefits everyone, the organization was a major player in the politics and socio-economic development of Alberta. This story documents the activities, the issues and the experiences of the women of this organization.

Even before they were formally invited into the membership of the United Farmers of Alberta (UFA) in 1913, farm women were busy organizing, not only to serve their own need to connect to other women, but also to respond to the pressing needs of their communities for facilities and services, needs that could only be addressed by cooperative action.

Leona Barritt, the first secretary-treasurer of the organization, documented the early years in a brief historical pamphlet in 1934. Eva Carter wrote a 45-page booklet entitled *Thirty Years of Progress*, outlining the UFWA history to 1944, drawing much of her material from Barritt's pamphlet. This book is the first to provide a detailed and comprehensive look at the ideas, the challenges and the accomplishments of organized farm women in Alberta who were members of United Farm Women and its successors. This publication recognizes 75 years of history of Women of Unifarm, from its first independent formation in 1916, to 1991.

The roots of the organization are deeper than the story of 1916, and so the early years, 1909-1916, will be discussed briefly in the first chapter of this volume. Chapter One documents the beginnings of farm women's cooperative efforts at the local level, when women identified a need to share their lives and their resources to improve their situation as farmers, mothers and wives. It also explains how the provincial organization was born, initially as an auxiliary of the United Farmers of Alberta in 1915, and then as an independent branch of the UFA in 1916. These organizational births occurred in an important historical context. Alberta was at a critical point in its development as a province, and in 1916, farmers were the trustees of Alberta's future. The significance of farmers' organizations and farm women's relationship to them will be discussed in the first chapter.

The second chapter, entitled "Farmers," looks at women's contributions to agriculture and their efforts to keep a class of people attached to the land. The ways in which women acted as farmers will be shared in this chapter: from politics to marketing strategies; from agricultural education to the development of shelter belts; from running farms single-handedly to financing farm expansion; from lobbying in Ottawa to planning policy with other provincial organizations in the Canadian Council of Agriculture. The ways in which women participated as farmers over the years are many and varied, and their commitment to the survival of the family farm is documented in Chapter Two.

Chapter Three, "Creating Communities," highlights the work performed by farm women to ensure people in rural communities could receive the education, health, social service and recreation opportunities they needed. The accomplishments of farm women in this area of endeavor are astonishing. They tackled many issues with dedication and skill, forcing politicians to be more responsive to the needs of Alberta citizens. Often the services were given directly by farm women, or organized by them. But their contribution was more than being "good neighbors," and this chapter will include a discussion of the significance of farm women's social service role to Alberta.

"Sharing fellowship" is the theme of Chapter Four. Interviews with farm women of several generations reveal that the United Farm Women of Alberta, and later Women of Unifarm, played a vital role in providing social contact and stimulation to farm women. Initially, many locals were formed because of the need to ameliorate the social isolation of farm life. And because the women extended their hospitality to their families and often to the whole community, the significance of the social events grew. In some communities, organized farm women were responsible for most of the year's important social events, and for fund-raising and catering for events shared with other community groups. For many farm women, the enduring friendships of many years are the most memorable outcomes of their membership in the organization.

Chapter Five provides an account of the commitment Alberta farm women made to the larger human community, at the national level and even beyond the borders of Canada. Perhaps less well known than their local and provincial activities, farm women's projects as "Citizens of the World" demonstrate that they had a broader vision of their place in the world and their ability to influence the course of world events through education, through national and international networking and through political activity. Chapter Five also traces some of the recurring issues and challenges farm women have faced throughout the 75 years. It presents the voices of today's farm women, as they talk about the future of farming and the future of their organization. This chapter will include some discussion on the evolving role of women in farming, and the new forms of cooperation and organization that farm women are initiating in response to a changing industry and a changing world.

Researching this book was a rewarding process, as many treasures in the form of speeches, briefs and letters were unearthed. Some of the best of farm women's writings are included in the final section of the book, for no one can adequately describe the eloquence, intelligence and commitment to service and to the welfare of humanity that are found in them. We thought you would enjoy them most in their original form.

This book is written from the perspective of the women who were part of the organization. Whether they were in leadership positions or were faithful members of a thriving local, women have important things to say about this organization and the role it played in their lives. While official documents help to frame the narrative, the real story lies in women's own words, as they remember their years of fellowship and toil, and as they reflect on the future of their organization. This history is not intended to be a definitive analysis of UFWA/Women of Unifarm and its place in provincial and national history. It is a book designed to describe, to remember and to celebrate. Perhaps a more detailed history could have been written if more sources were available. But many materials were discarded or destroyed, and many remain in private collections inaccessible to the author. Interview comments made by the women listed in the acknowledgements, who were interviewed during the period 1989-1991, are not included in the footnotes, but instead identified within the text. Only written documents or taped interviews of women no longer living are included in the notes.

The preparation of this history is a dream come true for many Alberta farm women. Some women have been an integral part of the organization for many years. Others wish to acknowledge the contributions of their mothers and grandmothers to the development of thriving rural communities. The difficulty in writing a history of this kind is that so much has to be left out. So many wonderful stories of fellowship, of personal achievement, of community spirit, and of good-hearted pranks and fun were shared in the process of compiling this book, and there was not room to include them all. Asking women to share their experiences and their perspectives demonstrated that we value and share these all too infrequently. Most women are not accustomed to being the primary subjects of their own conversations, or of valuing what they do and think as important and interesting. Yet somehow they know, in a very personal way, that what they have been part of as members of this organization has made a real difference, both in their own lives and in those of their neighbors. For this is truly a story of successful "grass-roots" organizing and "down-to-earth" caring for others. It is a story that took place quietly over many years in every corner of Alberta. It is a story that has been hidden from view by the bigger events of politics and economic change, yet it is integrally part of these bigger events, in some cases making them happen, while in others, responding to them. It is a story that women of this organization are proud to tell, and one that needs to be told. We hope you enjoy reading it.

Significant Dates

1905 Alberta becomes a province of Canada.
Alberta Farmers Association formed through an amalgamation of some locals of the Canadian Society of Equity and the Territorial Grain Growers Association.

1909 United Farmers of Alberta (UFA) is founded through an amalgamation of the Society of Equity and Alberta Farmers Association.

1912 UFA passes a convention resolution in support of women's suffrage.

1913 Women are invited to be equal members in the United Farmers of Alberta, constitutional amendment made to that effect.

1914 Canada enters the war; Alberta's population is close to 470 000, of whom two-thirds are farmers.
Women are invited to set up their own convention in conjunction with the UFA convention in Lethbridge; attendance is poor.

1915 Women are invited to Edmonton to consider organizing a provincial association: Women's Auxiliary of United Farmers of Alberta is created.
Forty women's locals are organized.
Southern Alberta farmers, including members of the UFA, form the Non-Partisan League, an alternative to traditional political parties.
Prohibition legislation is passed in Alberta, aided by the campaigning of farm men and women.

1916 The Women's Auxiliary is reorganized as the United Farm Women of Alberta, and the first provincial committees, Health, Education and Young People's Work, were struck.
Women obtain the right to vote in provincial and municipal elections in Alberta.

1917 The Non-Partisan League wins two seats in the legislature in the provincial election, filled by Louise McKinney and James Weir; women in Alberta vote for the first time.

1919 Junior Branch of the UFA, for farm girls and boys, is organized by UFWA; first Farm Young People's Week is held at the University of Alberta.
UFA/UFWA become political organizations, fielding candidates in federal election.

Interprovincial Council of Western Farm Women formed; they become affiliated with the Canadian Council of Agriculture nine months later.

1921 UFA/UFWA run candidates for the provincial election and win the majority of seats in the Legislature, becoming Alberta's first "group government"; the Non-Partisan League is subsumed by the political wing of the UFA.

1922 First radio stations open in Alberta: CJCA in Edmonton; CFAC and CFCN in Calgary.

1923 Prohibition is repealed.

1925 Egg and Poultry Pool is created (first proposed at 1924 Convention).

1928 First edition of the UFWA Cookbook is published.

1929 Persons Case is judged by the Privy Council, so that the word "persons" in the BNA Act includes members of the female sex, and they are therefore eligible to become members of the Senate.

1930 Farm Women's Week at Olds Agricultural College is initiated (originally called Farm Women's Rest Week).

1932 The Cooperative Commonwealth Federation is launched in Calgary, and the UFA and UFWA approve its Manifesto.

1936 Number of farmers at an all-time high of 99 732 in Alberta; coldest winter on record, hottest summer in thirty years with continuous hot dust storms.

1937 First rural chatauqua organized by the UFA/UFWA is held.
Worst disaster year: Lakes go dry, fences and road allowances disappear under dust. Six million acres of once cultivated land in the Palliser Triangle drift out of control.

1946 Farm population decreases rapidly as people move from farms to cities.

1949 Amalgamation of the United Farmers of Alberta and the Alberta Farmers' Union to form the Farmers; Union of Alberta. United Farm Women of Alberta vote to become the Farm Women's Union of Alberta.
The new magazine for Alberta farmers, "The Organized Farmer," publishes its first issue.

1959 Sod-turning at Goldeye Camp, near Nordegg. The camp was endorsed at the 1958 annual meeting; first youth program held there in 1961.

1970 Amalgamation of the Farmers' Union of Alberta and the Alberta Federation of Agriculture results in the formation of Unifarm; after discussion and a vote, the women's organization becomes " Women of Unifarm."

1978 The Government of Alberta introduces new Matrimonial Property Act to recognize that each partner has a legal right to an equal share of assets accumulated during the marriage. A major victory for farm women, who had been lobbying for decades on this issue. Act becomes effective January 1, 1979.

1989 A new edition (the eighth) of the cookbook is published. Over 100 000 copies of the cookbook (all editions) are sold. A history book committee is struck to develop the history of the United Farm Women of Alberta and its successors, Farm Women's Union of Alberta and Women of Unifarm.

1990 Women of Unifarm provincial convention in Red Deer votes to maintain the women's organization.

1991 A new constitution for Women of Unifarm is ratified.

1996 Women of Unifarm provincial convention in Lacombe votes to maintain Women of Unifarm as an independent and unaffiliated organization after the reorganization of Unifarm into the Wild Rose Agricultural Producers.

Alberta Farm Organizations

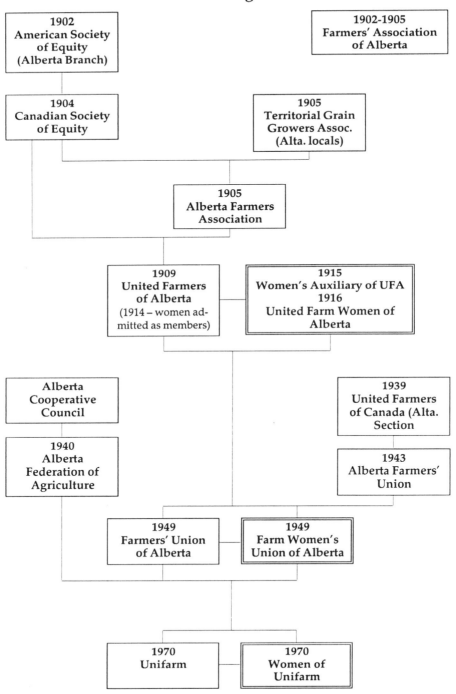

1902
American Society
of Equity
(Alberta Branch)

1902-1905
Farmers' Association
of Alberta

1904
Canadian Society
of Equity

1905
Territorial Grain
Growers Assoc.
(Alta. locals)

1905
Alberta Farmers
Association

1909
United Farmers
of Alberta
(1914 – women admitted as members)

1915
Women's Auxiliary of UFA
1916
United Farm Women of
Alberta

Alberta
Cooperative
Council

1939
United Farmers
of Canada (Alta.
Section

1940
Alberta
Federation of
Agriculture

1943
Alberta Farmers'
Union

1949
Farmers' Union
of Alberta

1949
Farm Women's
Union of Alberta

1970
Unifarm

1970
Women of
Unifarm

UNITED FARM WOMEN OF ALBERTA

All farm women are invited to join the United Farm Women of Alberta.

It aims to provide for the farm woman a social centre where she may meet her neighbors and enjoy an exchange of ideas in matters of interest.

It aims to make the farm woman herself the principal factor in her own development by giving her the opportunity for self-expression.

It aims to give the farm woman the means to extend her education in public affairs and matters that are of vital interest to every woman in Canada.

It aims to lead her to co-operate with all the people on all the farms for betterment of farm conditions.

It aims to study the rural school and arrive at an understanding as to what sort of education the rural child needs to equip him for the best, the most useful and the happiest life possible.

It aims to organize the farm women 100 per cent, and thus raise the standing of the farm women in the public esteem as a member of an organization able to demand a square deal.

The benefits to the farm woman in this Association are:

Education   Sociability   Co-operation
            Power through organization.

THE SECRETARY

Chapter One

Beginnings

The major catalyst that changed Alberta from a collection of farms, ranches and towns to a cohesive society was farm organizations. Picture a vast expanse of prairie, dotted with homesteads every few miles. In those little houses, unconnected to any other except by occasional mail and a rare neighborly visit, women labor from sunrise to midnight to feed and clothe their families, to make a comfortable home for them, and to help their partner eke out their modest existence from the soil. There is little time or opportunity to visit a neighbor woman, and most of the year, the roads, or lack of them, make a visit a day-long outing, a day that cannot be spared. They may request a neighbor's help in time of illness, catastrophe or birth, but for the most part, these farmers must be self-sufficient. There is no other way.

By 1905, two Alberta farm organizations, the Canadian Society of Equity and the Alberta Farmers' Association, had grown out of a need for economic cooperation among farmers, who recognized their relative powerlessness as independent producers, acting alone. By 1909, these two organizations understood the value of a united effort, and the United Farmers of Alberta (UFA) was born. This shift in perspective, from farming as an individual livelihood to farming as a shared industry, took place as farmers realized, through organizing, they could and should be politically influential. They were, after all, two thirds of Alberta's population, and their farms and ranches were the economic base on which Alberta's prosperity was to be built. As the UFA was being formed, Alberta was enjoying its greatest growth spurt of the century. In the decade between 1901 and 1911, the number of farms on the prairies increased from 55 000 to over 200 000, most of them being established in Saskatchewan and Alberta. Perhaps equally significant, the number of women in Alberta increased five-fold in the same decade.[1]

Alberta was dominated politically by the Liberal Party from 1905 to 1921. Despite a weak Conservative Opposition Party, the Liberal government was prudent enough to ensure the farmers' votes by responding to many of their demands. Still, farmers felt disadvantaged in the marketplace and believed that the interests of other economic sectors were placed before their own. Farming as an industry was affected by federal policy as much as it was by provincial legislation, and it was Alberta farmers' grievances with Ottawa as well as with the provincial legislators in Edmonton that united them in a common cause.

By 1912, the *Grain Growers' Guide,* the official organ of the United Farmers of Alberta and other prairie organizations, suggested to farm women that they organize too, for the betterment of themselves and the farming industry. A resolution was put forward at the 1912 UFA convention urging members to assist farm women to "organize provincially and locally along the lines of the UFA."[2] The new sense of community that arose out of a shared livelihood and common grievances had its effect on farm women too. They had concerns about the ways in which their lives and those of their families were affected by the policies and the indifference of governments. But they also recognized their own need to build community among themselves, to solve the problems of everyday living through cooperating with other farm women. They quite clearly saw the link between the economics of farming and the consequent conditions in which rural families lived, deprived of education, health and cultural services and opportunities. They characterized their organizational interests as simultaneously a partnership with the men in addressing the economic issues of farming, and as a separate caucus of women with a particular interest in the needs of women and children.

It was the need of farm women to connect socially with other farm women that lay at the roots of their organization, the United Farm Women of Alberta (UFWA). "Our farm women's organization grew primarily out of a pronounced need for some form of social intercourse," an early UFWA pamphlet proclaims. It continues: "The monotony and isolation of farm life, with its consequent restricted opportunities for recreation, development or service, was the despair of many a thinking woman."[3] As Irene Parlby wrote years later, farm women knew "organization . . . would mean mutual interests and less loneliness."[4] As early as 1903, women gathered in small groups in rural communities to share their lives and their concerns, to exchange ideas and to plan projects for their communities. These groups often met independently, without thought of a relationship to the United Farmers of Alberta, which began organizing local branches in 1909. In some cases, women organized a women's group or "auxiliary" in anticipation of the men in their community organizing a local branch of the UFA. By 1915, when the first provincial meeting of farm women was held, many local groups were in place with distinct names, and purposes and procedures of their own. Two examples of these groups are found in the communities of Alix and Edgerton.

Alix, which later became known as Local No. 1 of the United Farm Women of Alberta, is regarded as the testing ground for many of the principles and practices on which the UFWA would be developed at the local level. But the initial gatherings of farm women held in Alix did not have the UFA in mind. The first meeting was held in the spring of 1913 at the initiative of Jean Reed, a housekeeper for the Marryat family, Irene

Parlby's parents. At Jean Reed's invitation, Elizabeth Mitchell, a house guest she was entertaining, addressed a small group of women at the Anglican church in Alix. Her topic was country women's organizations. Elizabeth Mitchell was a writer from England who had been investigating the Homemakers of Saskatchewan and the fledgling Women's Institutes in Alberta. She talked about the benefits of organization for rural women. The women of Alix area decided to meet again, and to consider the possibility of forming an association that would meet on a regular basis. The following month, the Alix Country Women's Club was founded. At first, it was affiliated with the Women's Institutes, but when the Women's Auxiliary of the UFA was established the following year, it joined the Auxiliary, claiming that the new provincial organization was better suited to the needs of the Alix local.

In their first year together, the 26 founding members of the Alix Country Women's Club set up the first public library in the area, and developed a rest room and library facility in a vacant Methodist church. It was a place for farm women to rest with their children during visits to town, to drink tea and browse through books in the library. The books had been donated by people from all over the world in response to a letter written to a London newspaper by Irene Parlby. The club raised money for the Red Cross and assisted with provincial relief work. A club choir was organized, and monthly meetings focused on a variety of topics of interest to the members. As Irene Parlby later described them, the meetings "dealt at first, with purely womanly things, friendly gatherings over a cup of tea, and we laid the beginnings for some warm friendships which have lasted through the years."[5]

Edgerton was one of the women's locals that organized with an intent to affiliate with the UFA. A meeting of interested women was called at the home of Mrs. Spencer in June 1913, to consider the feasibility of forming an auxiliary to the UFA local. The general aims of the group were discussed. It was decided that it would be nondenominational, and was organized to give "a little social life to farmers' wives and daughters. It aims also to help along housekeeping lines, as well as help us forget our housekeeping for a while; help others as well as ourselves."[6] In August of 1913, Mrs. Spencer read a letter to the group from Mr. Woodbridge, Secretary of the UFA, who stated that the government and the UFA were trying to unite the Auxiliaries and the Institutes (Women's Institutes), which were practically the same. The Edgerton women's group chose to leave the naming of their society until this matter was decided. In October of 1913, after receiving correspondence from the University of Alberta concerning the travelling library, the women decided to hold a dance jointly with the UFA local (known as the McCafferty local) to raise funds to host the travelling library. Mr. Ottewell, of the Department of Extension at the university, lectured to the group several times in 1914, on direct legislation and on women's suffrage. When women were invited, by convention resolution, to become full members of the UFA in 1913,

South Edgerton Group. (photo courtesy Doreen Kennedy)

the men of the McCafferty UFA local invited the women to attend their meetings. The women extended a "vote of thanks" to the men "for their kind invitation" and continued to meet on their own.[7] The women's auxiliary sent a delegate to the first women's provincial meeting in 1915, and by April 15 they decided to take the steps to organize themselves formally as the McCafferty Auxiliary of the UFA. A variety of petitions, projects and topics absorbed the interest and energies of the group in 1915 and 1916. These included women's suffrage, the legal status of women, women's political equality in Manitoba, a public nursing scheme, a proposed hospital scheme, Red Cross fund raising, and an incident of child abuse in their community. The topic of feeding the threshers was thoroughly discussed in the spring of 1915, and a report detailing a plan of a cook car to feed threshers, and how to finance it, was developed and sent to the Secretary of the UFA.

Other women's groups, in communities throughout the province, were well in place by the time the push to organize at the provincial level was successful. They carried out similar activities to the ones in the Alix and Edgerton areas. Beginning in 1903, Alberta farm women had taken the initiative to organize, to meet their own needs for social contact and to identify and address the needs of their communities.

Where did the idea to develop a provincial association of women originate? Correspondence from the UFA Secretary-Treasurer, P. P. Woodbridge, to the executive of the Women's Section of the Saskatchewan Grain Growers suggests that the original organizing meeting held at the Edmonton UFA convention in January 1915 was at the initiative of the UFA central office.[8] While women had been invited to join the UFA in 1913, "on a basis of equality with the men,"[9] they had not been urged at that time to form a separate association, as a 1912 convention resolution had suggested to them.[10] Woodbridge's letter reveals two critical pieces of information. One is that the growth of women's local groups and the female membership in the UFA was sufficient to create a great deal of work for UFA head office, which Woodbridge largely did himself "with little or no assistance from the directors."[11] The second is that the UFA was interested in having a woman's provincial organization start the previous year, in 1914, along the lines of the Women Grain Growers in Saskatchewan, as an auxiliary to the men's organization. The *Grain Growers' Guide* assisted the UFA by advertising a separate women's meeting in conjunction with the UFA convention in Lethbridge in 1914. But it was poorly attended, and the intent to found a separate women's organization was not clear to all. According to Woodbridge "the movement failed simply for lack of a practical leader."[12] Whether the women of the UFWA would agree with his assessment of the situation remains in doubt, but whatever difference a year made, the historic founding convention did indeed take place in January 1915 at McDougall Church in Edmonton. Fifty-eight delegates from women's groups all over the province were in attendance, in addition to the speakers and organizers.

Several aspects of the founding meeting would have long-term implications for the farm women's organization. The first is that the initiative for organizing came from the central office of the UFA. However willing and ready some of the farm women were to take this step, the fact that the farm women's movement was created from the top down and not from the bottom up, where women themselves decided what kind of provincial association they needed, and when it was appropriate to develop one, would ultimately determine the character of the organization. It would become one that was highly centralized, with all local initiatives both funnelled through and directed by central office. This is evident in the development of a yearly program at the provincial level that most locals followed faithfully for many years. The program, accompanied by information bulletins beginning in 1920, included both the content and structure of monthly meetings, leaving local groups free to add their own preferences to the basic format. Equally important in this top-down style of organizing was the relationship of the women's section to the men's organization. Born of this larger and older organization, the relationship was always one of child to parent; in fact, references to the UFA and its successors by members of the women's organizations frequently use the words "parent body." Constitutional, financial policy-making and decision-making processes would institutionalize this parent-child relationship over time.

Another feature of the founding meeting was the inclusion of guest speakers and leaders from other provinces and from urban centres. Violet McNaughton was invited, as an experienced organizer of the Saskatchewan Grain Growers Women's Section (WSGG). The agricultural writer

Violet McNaughton. (Photo 412149 courtesy of the Saskatoon Public Library Local History Room.)

and women's editor of the *Grain Growers' Guide*, Francis Beynon, travelled from Winnipeg to the meeting. And participants in the historic gathering were treated to speeches by Nellie McClung, Dr. Alexander of the Extension Department at the University of Alberta, Miss Beynon and Miss Clendenan, another press woman. The program was put together very quickly, with little input from the farm women, many of whom travelled many miles to attend this first provincial women's meeting. Nellie McClung, whose childhood was spent on a farm, was well known for her idealized view of country life and her belief that farm life had religious and spiritual dimensions that contributed to building human character. All speakers emphasized the critical role farm women play in the development of a moral and educated citizenry, and in keeping the family on the farm. No doubt the farm women in attendance were inspired and supported by the knowledge that others felt that what they did was important, even vital, to the welfare of the province. More significantly though, these early connections with key urban and farm movement people would serve the organization well in the coming years. Violet NcNaughton and the other executive members of the WSGG would provide invaluable support to the executive members of the new Alberta organization, and the *Grain Growers' Guide* was adopted and well-utilized as the new organ of the women's branch. Dr. Alexander and the services of the Department of Extension were used by farm women to establish travelling libraries, as resources for research for information bulletins and for speakers for meetings and conventions. Nellie McClung would for many years inspire farm women and men alike with her convention addresses, and with her sincere promotion of farm life and of the responsibility of urban women to assist their rural sisters.

The final feature of this organizational meeting which had long-range effects was the decision to match the UFA in its membership requirements and create a class- or occupationally-based organization, that is, an organization of women farmers, not an organization of rural women. The presence of members of Women's Institute at the January 15 meeting made this an early issue that demanded resolution. The Women's Institute had already organized in many rural communities, and included in its local branches both town and farm women. The presence of an already viable country women's group, which had the enviable position of being funded by the provincial government, led those at the founding meeting into a heated debate and a difficult vote. There was some speculation that the government favored an amalgamation of all farm women's locals into Women's Institutes and the reality that many community groups were joining the Institutes is seen as the impetus behind central office's concern to get a provincial farm women's organization established as soon as possible. The result of the vote, however, was that two organizations would co-exist for rural women in Alberta: Women's Institutes, to include all women who wished to join, and the United Farm Women of Alberta, whose membership would be farm women, either the wives or daughters of farmers, or the sole operators of farms, who were actively engaged in farming. This decision made the two organizations distinct from each other as did other features. As Irene Parlby, second president of the UFWA, characterized it, Women's Institutes were more about housekeeping and had less interest and involvement in political affairs than the UFWA. The chief difference in the early years was that the Women's Institute was both organized by and funded by the provincial Department of Agriculture. It was this relationship that was responsible for its careful non-partisan stance. When the UFA formed Alberta's government in 1921, they soon proposed a change in the status of the Institutes, giving them five years to organize as a self-sustaining and independent organization, which Women's Institutes accomplished by 1928. At the local level, as the local branches of UFWA and Women's Institutes carried out their social, charity and lobbying activities, few differences in the two organizations were visible. Farm women were sometimes members of both organizations, and in small communities many projects were conducted on a cooperative basis by both organizations. Some of the local groups, one being Lea Park, switched their affiliations from Women's Institutes to the United Farm Women when the latter was organized. But antagonism between the two organizations would remain for many years at the provincial level, fed by the disparate financial arrangements under which each attempted to attract members and carry out their activities.

Another outcome of the decision of the women assembled to become a part of the UFA and to identify itself as a class-based organization was that the aims and priorities of the parent body became those of the new branch. So the new women's auxiliary adopted the educational and marketing foci of the UFA, and the lobbying role the UFA saw as critical to pressure government to enact legislation to improve the economic position and the living conditions of farmers. And initially, the women's auxiliary was dependent on the funds provided by the UFA to organize its provincial body. A grant of $100 and a personal donation by Rice

Sheppard of $ 61.70 were the start-up funds. With that modest amount, a great deal was achieved by the auxiliary in the first year. Forty locals were organized in various parts of the province, including women's local groups that had already been formed and were eager to join a provincial body. These locals represented a membership of over 700 women. As a UFWA pamphlet described that founding year:

> *This little body of women before their second convention, has plunged into Belgian relief work, were working for Dominion wide equal suffrage, had taken up the matter of travelling libraries with the Extension Department of the University of Alberta . . .*[13]

In 1916 the role of the auxiliary was redefined, as it attempted to form its own identity and purpose, separate from but still in support of the goals and programs of the UFA. There was pressure from one local, the Veteran Womens' Auxiliary formed in 1915, to change the name of the women's organization. On behalf of the Veteran Womens' Auxiliary, the men of Local Union 363 brought a resolution to the United Farmers convention to have the auxiliary renamed the United Farm Women of Alberta. It was passed on the first day of convention, giving the women time to consider other changes to their organization. Irene Parlby, who was elected president of the organization in 1916, and who was absent due to illness from the 1915 founding meeting in Edmonton, was also not satisfied with the auxiliary status of the women's organization. "I have never liked auxiliaries. It puts them in an inferior, tagged-on light."[14] Delegates responded to her suggestion that, in addition to their new name, they become an independent organization of women, setting their own agendas and formulating their own policies. Although Parlby engaged the support of the more progressive men on the UFA board for the proposed change of status, she wrote later that the UFA needed to be really pushed to accept this change:

> *The change did not occur as a natural sequence of events nor as a part of ordinary procedure but requires a great deal of persistence and patience on the part of the Auxiliary before the men realized the value of accepting the women's branch as an integral part of their organization with equal privileges.*[15]

And so the United Farm Women of Alberta (UFWA) was launched. Those "equal privileges" of which Parlby wrote were the right of the women to conduct their own affairs, to set up their own locals, and to have two (later three) of the UFWA executive members sit on the UFA Board. In addition, the women were able to convince the UFA board to increase their annual operating grant to $500. This did not mean that the women enjoyed independence in the formulation of policy. They still embraced the ideology and goals of the men's organization, and further, every resolution they passed at their own convention had to be brought to the UFA convention floor and passed by a majority vote before it could become the official policy of the organization. The women also had to have the men's support before they could put any proposal before government. The characterization of men's and women's work that evolved could be captured in two phrases, "trade and economics" and "social welfare." Men of the UFA were more focused on the economic issues affecting farming, and left the social issues concerning the quality of farm and rural life to the women's organization. The women required the men's support and involvement to tackle pressing issues like rural public health care and rural hospitals, rural schooling, services for the mentally infirm, child welfare and many other concerns. There is evidence that men's involvement was not always forthcoming. The men appeared to sign over the work on these issues to the women, in a way forcing them to specialize in these areas without a full appreciation of the economic implications of "women's work." But farm women were also producers, and did not confine their activities to only social issues. Cooperative marketing became an early and central focus of the UFWA. Where the separate missions of the UFA and UFWA would produce conflict between men's and women's points of view, as they did on the issues of birth control and matrimonial property, what was good for the UFA, as it embodied the unified front necessary for lobbying power, usually prevailed. The parent-child hierarchy of power was invoked, and women would sacrifice their goals and hopes for women on behalf of what men professed to be the goals of the farm community as a whole. Despite Parlby's insistence on a separate women's organization, she reinforced the supporting rather than equal role of women in the farm movement, suggesting women's locals serve to strengthen the men's organization, and as "the greatest force for bettering rural conditions that we possess."[16]

Parlby was a natural choice for president of the women's branch. She excelled in the areas most needed in leadership roles. She had vision of what the organization could be, and she was an eloquent speaker and a talented writer. She had already put her thoughts about the organization before the public in an article to the *Grain Growers' Guide* in April 1915:

> We are hoping great things from our new organization for farm women in the future. They will have many difficulties to combat, not the least being the apathy and indifference of the country women themselves, but there is a powerful leaven working amongst the more progressive and intelligent spirits on the farms, which is slowly but surely making itself felt, and if only every member of the U.F.A. and its Women's Auxiliary would come to regard themselves as missionaries to convert their neighbours to the great gospel of united effort, the speeding up of our organization and its future growth and power for good would be immense.[17]

Despite her own missionary zeal for the organization, Parlby had no ambition to be president of the women's auxiliary, nor to take a key role in its transformation. She had to be strongly persuaded to let her name stand for president at the 1916 convention, and it was Violet McNaughton and some of the Alberta press women who finally managed to convince her. Her nomination no doubt was inspired by the articulate and passionate address, however deferentially she began it, that she delivered to the women's convention. Entitled "Women's Place in the Nation," it is a speech about ideals. This was Parlby's forte as a leader. She was a person who liked to put ideas in front of people, to encourage them to reach further and think grander thoughts than their own daily struggles allowed. She was a spiritual leader, an idealist and a woman of great charm and talent, whose love of beauty and books, and whose education and life experiences distinguished her as a woman apart from many. As Mrs. Spencer, who worked with her as first vice president in 1916, wrote many years later, "Mrs. Parlby became president when farm women were timid and unsure of themselves and her confidence and poise encouraged and inspired many."[18] Her influence on the organization in the early years was great, and many of the ideas and procedures she brought to the organization would last for decades.

The task before the new president and her organization was daunting. As Irene Parlby would later describe it:

> I was elected President, with the gigantic and rather terrifying task of building up a Provincial organization, of evolving policies, and fixing a goal toward which to work. We had no money to work with, no prestige, and were to cooperate with a men's group which, while outwardly polite, did not at first realize the added strength which the women would lend to their own movement.[19]

As the letters of Irene Parlby would later reveal, the role of Violet McNaughton in nurturing the new organization and teaching its leaders was both critical and influential. Under McNaughton's direction and advice, many of the initial steps of organizing the UFWA, both centrally, in relationship to the men's organization and at the local level, were carried out in ways already tried in Saskatchewan in the Women's Section of the Saskatchewan Grain Growers. Parlby's letters to McNaughton immediately following her election as president reflect her need for guidance. She felt unsure about where and how to begin organizing women into a cohesive body, she worried about the lack of money to carry out the work, and about the resistance from the men of the UFA, many of whom did not share head office's interest in seeing a women's organization formed. Parlby named three goals for herself in that first year: to organize locals across the province and increase membership; to develop policies in areas of particular concern to women; and to provide inspiration through her leadership.

In her report to the *Grain Growers' Guide*, published February 9, 1916, Parlby identified the role of organized farm women as primarily educators in their farm homes and in the farm movement. The new organization began with both educational and social aims. The three identified in

1916 were: to increase and broaden the knowledge of farm women; to meet a pronounced need for social contact among farm women; and to cooperate with the UFA on all matters affecting the welfare of rural people. Thus, 1916 was another building year for the newly constituted organization. Irene Parlby and Leona Barritt conducted an organizing tour in southern Alberta in June 1916, visiting and helping to found nine locals.[20] There were no funds to cover the costs of their travels, and rough country roads made the trips uncomfortable and time-consuming. Women who were to receive their call to organize were often too busy farming, as World War I had taken many men away from their farms and families.

One of the new locals that was organized in 1916 was Carstairs. The minutes of that first Carstairs meeting record that 22 women assembled at the home of Mrs. Henry Wise Wood, wife of the president of the UFA. They heard an address by Mrs. Parlby. She "spoke for some forty minutes, setting forth some of the many problems, strictly rural in nature, which confronted farm women. She earnestly advised the farm women to have a strictly 'class' organization and to deal with these problems in cooperation with the United Farmers of Alberta. After the address, 15 of the ladies present decided to become members. They organized a local."[21]

During the first two years of the organization's life, both as an auxiliary and a branch of the UFA, locals were consumed by relief work in response to the drought in southern Alberta and by Red Cross work in aid of the war effort, as well as focusing people's attention on the need for "more adequate public health work, juster legislation for women and the betterment of our rural schools."[22] By the end of the founding year of the United Farm Women of Alberta, three important committees had been organized: health, education and young people's work, with a provincial convenor appointed for each to coordinate the work of the committee. In her annual report to the UFA convention, Irene Parlby mentioned her concern for the "comparatively slow progress in the growth of our numbers"[23] and appealed to the men of the UFA to encourage the development of women's local in their communities, or to have women join the UFA itself when women's locals were unfeasible. She acknowledged the more than double burden under which many farm women labored, as the war had created a shortage of men for farm work, and that "small wonder that she [the farm woman] has little energy left for Club work."[24] Irene Parlby wrote to the *Grain Growers' Guide* in August of 1916:

In the moments of despondency which must at times attack a mere mortal, when one wonders why one should trouble with these things or whether anything is worthwhile, it will only be necessary to think of those meetings of bright women, and their encouragement and sympathy, to feel one's courage rise again and to realize that, after all, if our organization can do nothing more than help to mould public opinion or help a lonely women here

*and there, it will at least have justified its being. That it has proved a help
to many women already by bringing broader interests and a wider horizon
to their lives we know for a fact from their own statements. . . . That many
members of the U.F.W. are doing some pretty good thinking we know, for
at different places we found them studying the tariff, political economy, in
spite of their busy lives and many home duties.*[25]

This passage tells the story of the exhaustion, of the discouragement
and of a feeling of renewal that organizers of the United Farm Women
of Alberta felt as they travelled throughout the province to give birth to
a new women's association. Leona Barritt later reflected in 1934 that "In
some quarters we were a joke, in some others we were wild-eyed suffrag-
ettes, while still others more sympathetic were absolutely amazed that
women coming from farms could possibly carry on as we did."[26] Parlby,
Barritt and the others who met together to form new locals united in a
common cause knew they were doing something important, for themselves and for their rural communities. Little did they realize that their
efforts would lead to exciting and productive years ahead for farm
women and the organization they had claimed as their own.

*UFWA Board, 1919. Back, left to right: Mrs. J.W. Field, Mrs. J. Dowler, Mrs.
M.J. Sears, Mrs. O.S. Welch, Mrs. Macguire, Mrs. Charles Henderson. Front,
left to right: Mrs. A.M. Postans, Mrs. J.F. Ross, **Mrs. W.H. (Irene) Parlby**,
Mrs. Paul Carr, Miss Mary W. Spiller.*

SOME AIMS OF THE U.F.W.A.

A prosperous agriculture.

A Canada united in its effort to secure uniformly high standards in educational and health services from coast to coast.

University facilities for rural boys and girls.

Adequate health services for every man, woman and child.

Intelligent, active public opinion to ensure permanent international peace.

Goodwill and friendliness in rural communities.

Co-operation between urban and rural dwellers.

To urge women to take the responsibility of citizenship.

To foster the principles and practice of true cooperation.

To find a better way of life for those who come after.

Chapter Two

Farmers

Women have always been working partners on family farms. Most farm and ranch operations are dependent on women's work for their success. The original organization of farm women was an expression of women's commitment to farming and to their desire to have full partnerships in the decision making that affected their lives. Women of the United Farm Women of Alberta always believed they had a role in "the economic problems confronting the farmer; problems which affected the wife and family no less than the husband and father."[1] This chapter focuses on how women described and fulfilled their vision of their partnership in the farming industry and the farm community over the 75 years and the difficulties they encountered in asserting their voice and their votes.

Special Role

Women of the United Farm Women and its successors, the FWUA and Women of Unifarm, considered themselves farmers first. They saw themselves working on behalf of farming people, both out of a sense that farmers were disadvantaged in many ways compared to urban citizens, and out of a belief that what was good for farmers was good for the whole province. They were concerned with representation of farmers' interests in the decision making and resource allocations of municipal and provincial governments. But they were equally concerned that women's perspectives and women's experiences be part of those processes. It is important to understand that in putting the farm identity first, women were not putting their needs as women second. Women of these organizations saw a specialized role for themselves within the farm community, as well as in the province as a whole, to work on those issues of particular concern to women, both as members of the industry of farming, with a particular investment in and view of farming distinct from the male view, and as nurturers of the next generation.

Farm women's advocacy on behalf of women and children was originally motivated by women's experiences in the homestead years. Focused mainly on the consequences of both isolation and lack of services such as health and education to farm families, women's social activism was a direct response to the conditions of homestead life. From personal experience and from awareness of their neighbors' experiences as well, women began to agitate for the most-needed services, professional maternity care, family health care and schools, long before a provincial organization was founded. Farm women's social activism arose not out of a need for an organization to find purposeful work, but out of the tragedies women lived and of the expectations women had for their new lives on the prairies. Women were integral to settlement. It was women who worked long days and nights to ensure the survival of their families on homesteads, and it was women who organized campaigns to build schools and obtain medical services. Dorothy Ottewell, who was associated with the organization for many years, described her view of the roots of farm women's activism:

> *Considering when it started we were slaves, [not] exactly slaves of course but as far as doing anything [we] certainly were. [We've] been pretty good pushing . . . the government for things in connection with education and health. Those were two of the most important ones in my memories.*

Many of the first generation of Alberta farm women were well-educated women, some of them arriving on the prairies to farm after working as professionally trained nurses, social workers and teachers in their home countries and cities. They had lived where medical care, schools, social institutions and other amenities were significantly better. They knew what they wanted and they knew what was possible for themselves and their families. Their understanding of the economics of farming, of the essential role of women in the enterprise, and of the power imbalance between men and women in families and in the industry all contributed to the particular advocacy and social activism role they strategized for themselves and their organization. The pressing needs for health care, education and social support systems and women's expertise and indispensability in the farm community combined powerfully to determine the ways in which women would choose to work on behalf of the farm community.

The early speeches of the organization's leaders reflect this duality, of women farmers who have determined they have a particular role to play on behalf of the farming community, while also recognizing that it was women's particular needs and concerns that they were advocating. They believed that those needs and concerns were inseparable from the viability of farming as a family-based industry, and the ability of each farm family to stay on the land. Their belief was grounded in their own knowledge of the farm's dependence on a woman's labor and expertise, and on the reality that successful farming was based on the full participation of all family members. A farm woman's interest in economics and world affairs was in fact an extension of her life as a farmer. As accountant and bookkeeper for the farm, she had to be aware of world prices, of freight rates and of the political events that affected the prices of farm produce. Mrs. J. B. Kidd, the UFWA provincial secretary in 1922, described it this way:

> *the interest the farm woman takes in public questions is a result of her environment, her life, herself. . . . the farmer does not know on the very hour he leaves home to market his drove of cattle or his pigs or his load of grain, what the returns for his labour will be. His wife soon comes to realize that the economic condition of the country has a very vital connection with the comforts of her home. . . . in the country even the initial step of organizing a school district rests with the handful of residents of that district. Thus the farmer's wife is brought in touch with the educational system of the province. All these factors in the lives of rural people tend to make the farm woman look below the surface of every question, get down to fundamentals, seek out the elusive "Why"?[2]*

Farm women have never seen the economics of farming as strictly men's concern, yet at the same time they recognize that men's power has always been based on their control of farm economics, on individual farms, in farm organizations and in provincial politics. Farm women recognized early in their organizational history and repeatedly throughout the years of partnership with the men's organization that access to power meant involvement in economic decision making as well as social policy formation, and access required knowledge. As Betty Pedersen, president of the women's organization from 1969 to 1975, explained it, the ability to participate in the economic discussions with men at the FUA executive table was an important way to command respect both for the women's president and the organization she represented. Irene Parlby understood the need for women to be represented on the executive of the UFA. The right of the women's organization to have its president on the executive of the larger organization began in 1916 and lasted until the latter became Unifarm in 1970. Then, as Betty Pedersen explained, women had to be elected to the executive of Unifarm and as president of Women of Unifarm, she made sure she was elected. The president and vice-president of Women of Unifarm were still automatically members of the Board of Directors of Unifarm, on an equal basis with other commodity group members. Even so, it was clear that throughout the years women had to assert their right to participate in the economic decision-making processes of the farming industry. Pedersen remembered: "You know, there was one director who said to me, when I was on the FUA executive, 'What are you doing here? You should be home in your kitchen cooking your husband's supper.'"

Certainly each generation of women experienced blatant sex discrimination in the farm organization of which they were a vital part. Irene Parlby talked about the challenge the women faced. "There is, of course, still discrimination, conscious and unconscious. . . . Eternal vigilance is needed if women are not to lose some of the things they have gained."[3] Many women expressed the frustration of having to "prove" themselves

The Board of Directors, United Farmers of Alberta, 1948. The women who represented the UFWA that year: Standing, left to right – Mrs. W.C. Taylor, 2nd Vice-President, Wainwright; Miss E. Birch, Secretary; Miss M. Coupland, 1st Vice-President, Lethbridge. Seated – Mrs. M. E. Lowe, President, Namao. (photo by E.W. Cadman)

capable and credible to the men of the organization. This task became more difficult as the men's and women's organizations became increasingly specialized and as their activities, particularly the annual conventions, became separate, beginning in 1970. Mary Newton reflected in 1990 on the original decision to form a separate women's organization: "I think I would like to have been there when it started. And I would never have seen an organization develop like this. I cannot see this division, really, why it ever went this way. . . . I think women just didn't play the aggressive role that now you can. But hindsight is always better than foresight."

The issue of separate organizations versus one organization for men and women, both locally and provincially, has been the source of much debate and speculation throughout the history of the organizations. It has also been divisive, as members of both sexes tried to come to terms with men's and women's needs and roles in farming. Women originally framed the separate interests approach in a discussion of unity and mutuality:

We believe this spirit of mutual trust and helpfulness is the backbone of the farmer's organization. It is impossible to separate the interests of farm women from the interests of farm men. Neither can be complete without the other. It is a matter of pride to the UFWA that it is not working independently and alone for better conditions for women only; but it is working shoulder to shoulder with the farm men of the province for justice and equity for all classes. It is true that the functioning of the two are slightly different, just as women's work in the home is different from man's work; but also as in the home, when once a definite purpose is decided upon, they work together for that cause with harmony, unity and power.[4]

Many locals worked out their own patterns of working together with the men, and their working definitions of mutuality and unity. At the provincial board level, a working partnership was always present. However, in the mid-1970s, the women's and men's organizations began to move further away from each other. This was supported by a constitutional change that no longer allowed the women's president to sit on the executive of Unifarm, and by a practice of holding separate conventions. Louise Johnston expressed her belief that women had lost some influence in the farmers' organizational voice because of the changes made at the formation of Unifarm in 1970. Mabel Barker was quite outspoken about the change in 1970 to separate conventions for men and women:

The biggest mistake that we made at the time was when we stopped having joint meetings, joint conventions. . . . Now when we [were] in the same building at the same time whenever there was anything of particular interest to the women we adjourned our meeting and went in with the men. We did that all the time so that we got in on all of those studies that were of interest. And when they stopped . . . then we lost out on that.

Joint meetings were historically the pattern for some locals, and until 1970, joint provincial conventions were held, with some separate and some shared sessions. Margaret House, a member since 1931, remembered:

I think it was sad that many women did not take charge of their lives or let the men do it all. . . . I have to thank my sister-in-law, who was a director then, and my husband for getting me to the conventions . . . and although men made the big decisions, I always knew what was going on and I think the women that went to the meetings knew what went on and then we would meet jointly with the men when it wasn't quite a busy time. But we really knew what was going on and tried to better ourselves. . . . And a little later I can remember, when I would be learning about the surface rights. . . a man from head office came and spoke and that was when people realized the oil industry was going to affect us. And the women were certainly the forerunners in that. It seemed in many instances the women were the forerunners, but yet they worked with the men on it, but they kept the organization going. . . . They kept them informed and kept them having their meetings, encouraging them too. And we'd have joint meetings to hear the convention reports but then we'd break off into our own sessions, which interested us the most. . . . I felt that the woman's organization [was] in a much better position than they are today with the main organization. We sat side by side and it seemed like we were considered a lot more than they are at conventions today.

Speaking of her days on the executive of the FWUA, Florence Scissons said:

The men didn't get away with too much because we had our own ideas and we tried to be realistic. We were also quite capable of sticking up for ourselves, so there really was no problem, but we worked with the men too on whatever projects there were and pulled our weight as well as we could. It was interesting at the conventions, the conventions were joint at the time, and so the women would draw up their program and the FUA would do the same. Mind you there were three women on the FUA Board anyway, so we would all be together for the session when we drew up the programs for the convention. But it was quite interesting that [when] we had what we thought was a good speaker, then the men would ask that we have that [speaker] for the joint session too so that they could hear too. So it was good and a couple of the days were joint so that we could hear what we wanted to hear in the other session. It worked out quite well really. And we shared our expertise I think and our ideas and also our energies because it took a lot of work, a lot of time.

Some locals held joint meetings of men and women and found there were real advantages. Jean Rose, of the Grandmeadow local, offered her experience:

Well, I certainly see a big change in our Women of Unifarm [local] from what it was before, because we meet with the men now and I doubt that we would ever organize again as a separate women's group. I just think that meeting with the men has given us a lot of different interests and really broadened our interests in the farm itself that we didn't have before. We used to be kind of a do-gooder kind of group, you know, and we would always be putting on things to raise money to give to something else, but now I think we're more concerned with having a speaker out, having a film, having a debate. I really think our interests have switched over to the farm economy and what can be done about it. I think women are considered more a partner, a business partner, than we used to be in those days. And I think in a way we may have brought the men over our way too a little bit, so they're thinking more of the environment and about what happens at the school and things like the safety of the area and so on. There are a lot of ways in which we exchanged ideas which I think is good.

By insisting on women's presence at the UFA executive table, Irene Parlby set the tone for the historical practice of organized farm women blending the economic concerns of farmers with a campaign for improved conditions in rural communities and on farms. Farm women's activities throughout the years have demonstrated their broader vision of the interrelationship between farming as a business and farming as a way of life.

As one farm woman in the One Tree local near Brooks stated:

I'm afraid that women in this district did far more toward the farm organization than any of the men ever did. I think the women operate differently than the men though. I think they look at things a little differently and they look at things more socially and more family oriented . . . and even though you're interested in the farm operation, you still look at it from a family point of view rather than necessarily just a business point of view. You can have both.

Jean Culler, who joined Unifarm in 1973, sees the differences in men's and women's perspectives this way:

I think that the women have more sensitivity on a lot of issues that the men maybe don't have time for nor the same type of concerns. Not all of them, . . . but a majority of men wouldn't be interested in certain issues that the women would be [that] are just as important. Your moral issues and family issues are every bit as important as your economical issues. And yet the women are still free to go into Unifarm and that is why I encourage women to do that . . . they can also have that input into regular Unifarm, and I always like to encourage them to attend the regular Unifarm meetings as well.

Farm women's dual vision found expression in a variety of forms. One was in educational messages and activities for farm women. Some are visible in speeches at conferences and in their writing, while others took place in discussion groups at locals, with information provided by central office or other sources. Leona Barritt, writing in 1916, described her vision of farm women's role:

There is no restriction in our organization on what we may discuss; free wheat and better agricultural credit are not taboo. Our interests extend beyond the four walls we call home, to the larger home which is our Province, our fair Dominion and the World. . . .[5]

Other expressions of the dual vision were found in women's active support of agricultural lobby campaigns launched by the men, and by their own activities in researching, lobbying and organizing on a wide variety of agricultural concerns, at both the provincial and the local levels.

Locals

The success of the organization as a voice for farmers' concerns lay in its structure. The base of the structure, the locals in each small community, brought a dozen or more farm families together. As long as the locals remained numerous and strong, the organization remained an effective voice for farmers. Once the locals began to disappear, the boards of both the men's and women's organizations lacked the kind of input and support that had made them so successful and so respected as a lobby group in previous years. The structure worked in this way: issues would be discussed at local meetings, often resulting in further research and consultation among local farmers. In many cases, but not all, the issue would emerge as a resolution, which would be passed on to a sub-district convention, and then to a district convention and then finally to the provincial convention. Sub-districts and districts were replaced by regions as organizational changes were made at the founding of Unifarm, so that the flow would be from a local to regional convention and on to the provincial convention. Miriam Galloway expressed the local's educational and political role this way:

> It gave farmers a common goal on many issues, so often a farmer might think he's the only one with a certain problem and then when you get together and start talking you realize, maybe this is a problem we can do something about. . . . it provides information to the grass roots from the provincial organization and from the department of agriculture. And it makes us aware of problems in other areas. Some of them similar and some of them very different. . . . And of course, if the executive lobbies with the provincial government to change laws, that's really a very important part of our cause. And the farm women present a unified front representing agricultural producers when we make requests. . . .

If the issue was urgent it would be taken to a provincial board meeting, and not be held until provincial convention. Members of the women's organizations were proud to describe this "democratic" structure of policy making and lobbying on which the farm organization was built. At the same time, however, the flow of information from central office, including the monthly programs and bulletins, often influenced the issues on which women would spend their time and energy. So the flow of influence was essentially two-way, from the leadership at the top of the organization, represented by the provincial board, and from the farm women in the locals, who represented the particular interests of their communities and their own farms. As Hazel Braithwaite, president of the FWUA from 1958 to 1963, explained it:

> We covered a pretty large field. Each year's resolutions kept us well-informed as to community needs and the changes in policy that were necessary. Our group was aware of the laws and advocated many worthwhile things before the government seemed to be aware of the situation.[6]

Within the organization, there was plenty of room for differences of opinion, as many heated debates on a wide variety of issues evidence. Perhaps the most divisive issue between women was the area of reproductive rights, and women's control over reproduction, as it found expression in each generation in discussions of midwifery, birth control, sex education and family planning, and abortion. There was some tolerance for differing political ideologies, and memberships of many locals were split between political choices of the day. Most of these differences were accommodated in an atmosphere of relative tolerance over the years. The exceptions were, of course, when the organization of farmers themselves became political and fielded candidates in federal and provincial elections. All organization members were expected to support the farmer candidates and the political goals of the organization. Tension grew as opposition to the farmers' group government mounted over the years. The greatest political tensions were felt in the approach to the 1935 election, when the Social Credit movement swept the United Farmer's government from power. The CCF had attracted many farmers interested in a cooperative social philosophy. Despite the United Farmers' government investigations and rejection of Social Credit ideals, many farmers were drawn to the promises of Aberhart and his Social Credit movement. Women claim the bitterness felt from the "defection" of many farmers' government supporters to the Social Credit cause was deep and lasting in many communities, and affected the strength and effectiveness of UFWA locals. With reorganization and union with the Alberta Farmers' Union in 1949, the organization seemed to be recharged at the local level, and the 1950s brought an increase in new locals being organized, and a strengthening of the farmers' caucus to a membership of 30 000, 19 000 of them being women.

Political Training

The local, made up of a group of neighbors and friends, was the farmers' greatest strength and also their greatest weakness in the structure of the organization. Throughout its 75 years, the health of the organization has been measured not only by its number of members, but also by the number of "active" locals that were operating, and feeding into the higher levels of the organization. It was at the local level that women were first invited to join the organization. It was where they got their first experiences at public speaking and at chairing meetings, their first exposures to the issues, and to the structures of both the organization and the government in which the issues were to be addressed. As a UFWA pamphlet described it:

Here was a great Provincial training school for citizenship, where women stood on their feet and expressed opinions, haltingly at first, perhaps[,] but as time went on they became convinced that these opinions counted, for one by one they saw them incorporated into laws, laws that would endure and be an aid to women of future generations.[7]

Some prairie farm women had little experience with public speaking, and with the formal aspects of meetings. Nellie Peterson expressed her belief that farm women wanted to have a group separate from the men's, because many of the women "were hesitant to get up and speak on any subject while the men were there. They felt that the men would think they expressed themselves better than the women did, whether they did or didn't." She explained:

> So they felt more at ease with just the women. They also took their children with them, and women don't feel the same way about a baby when it cries in the middle of a meeting; they don't feel the same way about a toddler running up and down the aisle. It doesn't disturb the business they're doing. Men become absolutely useless, almost, when a child disturbs their conferences. So it was easier for women to do it that way. And then they had concerns that, while they were men's concerns, too, nonetheless those concerns impinged themselves more vitally on the women than they did on the men.[8]

Jacqueline Galloway, a young farm woman who joined the organization in 1984, explained women's personal growth in the organization as a positive outcome of both local and provincial board participation:

> I can't tell you how many times women, particularly on the board, have said to me, "I can't believe I'm doing this," whether it was giving a speech, or organizing something or chairing a session. Five years ago or two years ago, if you'd had told me I was doing this I would have never believed it. . . . I have seen Women of Unifarm do that for women, really help them blossom and discover things about themselves.

In many communities, the farm women's organization was the only opportunity for women to be exposed to social activism of any kind. Its power lay in the direct relationship of that activism to their daily lives. If an issue did not directly affect their farm and their family, then the sense of being useful and of contributing to the larger picture beyond her farm gate certainly affected the farm woman herself; it gave her a sense of purpose and of community. As Lena Scraba described it, "it's a good feeling to be able to say, well, I tried my best to help our fellow farmers." So it was a sense of need at both the communal level, fighting for what farmers needed, and at the personal level, the reward of being involved and useful, that motivated women to be active in their local farm women's group.

Perhaps the greatest barriers women experienced to participation in the organization were practical ones. Problems such as bad or impassable roads and trails, harsh weather conditions, and the requirement to attend to farm and family responsibilities worked against many keen women. For them, the lost opportunity to attend a meeting was a considerable disappointment. According to some women, some of these barriers were challenges that men overcame more easily. Men were not as tied down by young children, meal times and food production activities. Louise Johnston recalls her first invitation to attend a meeting:

> It was decided to start an organization in our particular community. One of the girls came up on horseback and told me they were going to organize on a certain day and I was canning beef and I had to can it that day so I just couldn't go to the first meeting. . . . I don't think I missed any more meetings after that.

Many women describe how difficult it was to get to meetings when their children were young. Some claim they simply did not participate in anything off the farm until their children were grown; it was impossible to leave them, and too difficult to take them along. More than anything the demands of the farm operation prohibited women's participation. When farms began to suffer because of women's absences or activity in the organization, women cut back or retired from their organizational work. This was a long-standing problem for farm women, from the days of Irene Parlby, who worried about the effect of her frequent absences from home in the early years of her presidency, to the decisions made by board members Mary Newton and Margaret Blanchard in the 1980s to reduce their organizational work because of the deleterious effects it was having on their farm operations. Some women were able to compensate the farm for their absences by hiring help. Other women were not in a financial position to use this solution. When asked about the barriers women had to overcome, Jean Buit explained:

> Well probably the hardest thing is your own self, your own family and your own farm. . . . you're just torn every which way. I think that's the hardest thing. And there's a lot of complaints that there's not enough women on boards and things. And many times the men are being blamed for that. But I don't go for that. I think women themselves know how far they can go and I know one time we had our whole board around the table, and we went around it and I don't think there was anyone there who was willing to spread themselves any thinner. They know where their limit is . . . because there is a limit as to what you can do.

The source of women's success in maintaining a provincial organization that continued to push the women's agenda was their talent for establishing and keeping active locals. In comparison, the men were not as effective as local organizers, and in many communities they relied on the women to keep locals socially and politically active. An FUA Director's report to the 1951 convention expressed appreciation to the women for their organizational work at the local level:

For the untiring efforts of the many women who work side by side with the men in the FUA locals we say "thank you". For in many isolated districts it is the women who keep the locals functioning.[9]

Women learned early the power of organization as local improvements led to bigger projects, often on a provincial scale. The essential lesson was the understanding and use of the power of numbers in the political sphere. As Myrtle Ward described it:

You don't write to the Government and say: "This area is destitute of nursing care" unless you can say: "We have 40 women at Milo and we have 55 women at Arrowwood and they are asking you to set up a program so that people will have nursing care." You couldn't solve all the world's problems on your own. You just didn't get anywhere unless you had an organization behind you.[10]

Locals Focus on Agriculture

The minutes of local meetings over the years indicate the diverse issues women discussed, many of which were developed into resolutions for presentation to government, or acted upon immediately at the local level. (Those activities that dealt with the social welfare aspects of community building will be discussed in more detail in the following chapter.) Throughout the years women continued to focus on the farm industry and developed resolutions on land use; on market issues, such as prices, distribution, packaging and the development of cooperatives; on the use of chemicals; on shelter belt development and fuel taxes; and on awareness programs about agriculture, particularly for young people in school curricula. In more recent years, studies and resolutions on stress in farm families, intergenerational transfer of farm businesses, wills and estate planning, and farm safety demonstrate a focus on the health of the family farm as an economic unit. As Irene Wagstaff, a recent executive member, said, "women are interested in economics just as much as men." And Joyce Templeton expressed the same sentiment in 1989 when she declared, "Women who get involved in agriculture are interested in every aspect of the farm operation, just like the men."

Although many of the agricultural issues with which women concerned themselves were raised by the provincial board through bulletins or other communications, or brought to the women's groups in the form of requests for support by the men of the organization, the local groups identified and pursued a wide variety of agricultural issues of particular interest to them. Included here are just a sample of local activities through the years.

The Roseleaf (later Blindman) local reported that "debates on a wide range of subjects were popular . . . subjects included tariffs against U.S. goods."[11] In 1920 the Asker local established a relief fund for the drought area of the province, and discussed farming techniques, including chicken dressing and marketing. The Ridgewood local held discussions on the egg pool, the wheat pool and on poultry raising in 1925. In 1929 the Ridgewood local sponsored an agricultural pests contest for children, lobbied about the price of binder twine and discussed papers presented on agriculture and turkey raising. In 1930 the Eastburg local forwarded a resolution on marketing potatoes and vegetables, requesting that a pool be established in Edmonton to handle the sale of these agricultural products. The Pembina local had a beekeeping presentation at one of their meetings in 1937. A paper on "women and agriculture" was read and discussed at the March 1940 meeting of the Readymade local. In 1940 the Clover Bar local hosted a presentation by Mrs. Scroten, a director of the Federal Board of the Canadian Federation of Agriculture, who spoke about rail grading and hogs, and the marketing of wheat. The Ridgewood local discussed the wheat reduction plan and a central office bulletin on farm gardens in 1941. In 1943 they discussed a circular on the low prices of wheat and bacon and sent telegrams to Ottawa about those concerns. The Clover Bar local sent telegrams to Ottawa in 1947 regarding low livestock prices. The Clyde FWUA formulated resolutions for the 1949 convention on compulsory herd testing by government, and government inspection of premises where cream was handled. In the same year, the Arrowwood local developed resolutions concerning the simplification of farmers' income tax returns, and financial assistance in Canada to youth who want to become established farmers. In 1950 the Readymade local held a joint meeting with the men to discuss grasshopper control. That same year, the Vulcan local proposed that a rural electrification association be set up. In 1956 the Freedom FWUA local sent a protest to The Dairy Pool head office in Edmonton about the poor grades and short weight measures on cream sent to the local creamery at Barrhead. In 1958 the Three Hills local set up a committee to study the price spreads on white bread and eggs. In the same year, the Burnt Lake local discussed resolutions for setting up a regional board for employment of women as emergency farm help.

The 1970s demonstrated a strong concern for legal and environmental issues. Many locals focused again on women's property rights as the Murdoch case became public. The Jefferson local raised concerns about illegal buying and selling of machines and also engaged speakers from the Department of the Environment to speak about air, water and soil pollution. Land use became a major focus in the 1970s as an important environmental concern and economic issue. In 1978 the Region One conference included sessions on "the proper way to keep farm records, filing systems, the best farm accounts books to use, etc."[12] and on wills and estates. And farm family stress came to the fore as a major issue, as locals participated in a major study, "Coping with Stress on the Farm," in 1978, and discussed the relationship between stress and farm accidents in Alberta. Fourteen workshops on farming and stress were conducted throughout the province with resource persons supplied by the Department of Agriculture. In the 1980s, the concerns were on farming being treated as a business, the ways in which farm women are perceived by others, farm safety and land use, including preservation of natural habitat. A major study of "Perceptions of Farm Women" was launched in 1984. Almost 1000 rural residents completed a 12 question survey, attempting to determine the perceptions of farm women by men and women who were not involved in farming as well as by those who were. Jacquie Jevne, of Millet, who conducted the study, observed " that our stereotypes affect the way in which we communicate with one another and therefore court interpersonal relationships. If this study raises our awareness of how we perceive ourselves and how others perceive us, and therefore cause us to consider our interaction with others it will have served its purpose and met the defined objectives."[13]

The Westlock local developed a resolution in 1986 asking the Alberta government to "direct AGT" to give private phone line service to the farmers "immediately."[14] The resolution stated that a farm is a business and that many farmers need a computer to run their businesses. The Sunniebend local passed a resolution that asked that landowners be encouraged to leave a minimum of one acre per quarter-section of land with trees on it, "either along a fence line or as a clump"[15] and that tax concessions be available to landowners who follow this practice.

For years, many locals supported youth programs, both the "Junior UFA," founded in 1919, and the 4-H clubs. The goal of these efforts was to interest and support young people in pursuing agricultural careers. It is important to note that although both men and women were concerned about keeping their children involved in farming, it was the women who organized the young farmers, and lobbied for government support of agricultural clubs like 4-H and for agricultural training.

Provincial Board and Regional Initiatives

At the provincial level, one of the outstanding accomplishments by women was the formation of an Egg and Poultry Pool in 1925. The pool was first advocated at the 1924 convention, and Mrs. Wyman served as its first secretary-treasurer. Susan Gunn, in her presidential address to the UFA convention, described the significance of this cooperative marketing venture:

> We are proud that in this department, as in others, Alberta has taken the lead, inasmuch as our Egg and Poultry Pool is the first in the Dominion to function on a province-wide basis. . . . Our farm women are becoming genuinely interested and keenly alive to its possibilities as a source of increased revenue for the farm home. We feel that just as our men made history in 1923 with the inception of the Wheat Pool, so have our women this year, through the organization of the Egg and Poultry Pool, made possible a new era of increased interest and prosperity to rural homes.[16]

As Myrtle Ward would later point out, however, once the Egg and Poultry Pool was established and began to make money, the men took it over. She claimed this followed a familiar pattern of women doing the formative work while men enjoyed the successes produced by women's labor.

Other important farm issues commanded the women's attention at the provincial board level. The UFWA worked for the revisions to the income tax law and the New Estate Tax Act, and addressed the need for credit legislation for young farmers. In 1924, UFWA President Susan Gunn was appointed to the Central and Northern Alberta Land Settlement Association.

In 1930, 34 United Farm Women attended the first farm women's week, organized by Isabel Townsend at the Agricultural College at Olds. A significant part of the program for the week, besides the welcome rest and social activities, was on farming techniques. Labor-saving procedures, fruit growing, hogs and landscaping were discussed. For 56 years until the program ended in 1986, groups of farm women came together for farm women's weeks to discuss agricultural concerns, learn new farm and homemaking skills, and to have a break and some fun away from the routine of the farm.

For some years, many locals obtained farm supplies through central office in Calgary, using the advantage of collective buying to obtain lower prices. This was the beginning of the co-op stores and the UFA cooperative. Women were integral to the success of this method, and its growth into more formal organizations. The board focused its attention in the 1920s and 1930s on the establishment of co-ops and the issue of grading poultry. In 1929 Mrs. Elridge gave an address on marketing at the annual UFWA conference. Mrs. Ross, president of the UFWA in 1933, centred her address to a regional conference on "the farm in the capitalist economy" and "cooperative marketing." Several of the 1930 bulletins issued to the locals featured information on beautification and horticulture, and a questionnaire was circulated on the attitude of banks toward lending money to farmers. In 1932, the women organized a campaign to send seeds to one of the dried out districts of Alberta.

In 1940, the UFWA convention featured a speaker, Ronald Pye, who informed the women of the work of the Canadian Federation of Agriculture and its aims and plans. In 1944 the Minister of Agriculture asked members of the UFWA to sit on an administrative board for the provincial schools of agriculture. The war years brought new issues to agricultural producers. Bulletins on war time prices and trades and then later on post-war agriculture were issued by the executive to the locals. These were followed in the post-war years by bulletins on the International Federation of Agriculture in 1947, and information bulletins on cooperation and a report on the Alberta Federation of Agriculture in 1948. In 1949 the issue of freight rates, and in particular the Crow's Nest Pass rate,

absorbed women's interests and energies as discussion resulted in resolutions to be forwarded to the board and on to the federal government.

Rural telephone lines had been installed and serviced by farmers. In
1951, farm women forwarded a resolution to the government asking
them to assume responsibility for rural telephone lines. In 1957, the
women presented a brief to the provincial government requesting loans
to young farmers and discussing the egg marketing issue. In the same
year, they met with the Minister of Finance to discuss concerns they had
identified about the status of women with respect to the Succession
Duties Act. They also met with the Canadian Association of Consumers
to clear up a "misunderstanding" and help make the Association "better
aware of farm problems" so that the CAC would be "more careful in
preparing resolutions on agricultural produce" and "have the producer
in mind a little more than previously."[17] In 1958 The Southwest District
7 conference formulated resolutions on the Intestate Act, farm labor
problems, on chemicals in foods and on deficiency payments. In May of
1958 the FWUA presented briefs to the Stewart Commission. Each brief
dealt with aspects of the farm economy affected by increasing spreads in
relation to what they produced and sold. In 1962, resolutions were sent
to the premier asking that the province open an irrigation agricultural
school to suit the needs of southern Alberta, and that a veterinary college
also be established. In 1963, in a brief to Cabinet, they requested that the
Intestate Succession Act be amended to increase the widow's entitlement
up to $30 000 before any distribution is made to the children. In 1968 the
FWUA presented a brief to the Royal Commission on the Status of
Women, highlighting issues like taxation laws that prohibited farm
women from claiming wages for their farm work, and requesting that
farm women be entitled to at least half of the estate upon widowhood.

In a brief to the Provincial Cabinet in 1973, women requested precau-
tionary tactics by the Department of Environment when spraying crops,
as they were concerned about environmental pollution. Also in 1973,
Women of Unifarm became involved in the Irene Murdoch case, rallying
support for the campaign to improve the property rights of farm women,
and thus fairer property settlements upon divorce. This campaign was
waged for five years, resulting in 1978 in the passing of new legislation
(effective January 1, 1979) that acknowledged the equal contributions of
women to farming by an equal sharing of all assets gained after marriage.
In 1973 farm women also called for a complete review of the Farm
Training Program. In 1975 Women of Unifarm sent a brief to the Alberta
Land Use Forum expressing their concerns that agricultural land be
reserved for agriculture and that a classification of fertile soil be developed and used as a guideline in determining land use. The brief also
addressed the practice of stripping land to provide topsoil for urban
development projects. In the fall of 1975, Women of Unifarm held surface
rights meetings throughout the province. In 1977, in a report entitled

"Concerns of Rural Women," Women of Unifarm focused on land ownership and use, and farm economics as major concerns.

The 1980s saw a resurgence of the child care issue. Women were needed more often to work as farm laborers on their own farms, and many also worked off the farm, and the question of how to care for the children resurfaced. A Region Three resolution drafted in 1986 asked government to set up a separate program for farm women, allowing them to apply for special assistance to help with the costs of child care in rural communities. Women of Unifarm submitted a brief to the Special Committee on Child Care in June 1986, offering five suggestions to help meet the child care needs of farm families. They identified the particular needs for child care on a seasonal basis, when mothers were tied up with the busy seasons on the farm. The growing concern over the safety of children on farms and the high cost and scarce availability of child care led women to join with other farm women's organizations in a special child care project in the early 1990s.

Region 4 conference sessions in 1982 focused on the development of contracts in plain language and on decision making. In 1984 Region 4 included a presentation on vegetable gardens by a horticulturalist at their annual conference. Region 4 Women of Unifarm organized the Agricultural Societies Bench Show in 1984. In their submission to the caucus committee on agriculture in 1986, Women of Unifarm expressed concern about the increasing use of chemicals in farm operations, and "the danger of the careless use of the chemical as well as the disposal of containers." In 1988 farm women identified to the Premier's Commission of Health Care a need for research into agricultural chemical use and its effects on the health of farmers. And in their submission the same year to the Agriculture and Rural Affairs Committee, they expanded on their concerns about pesticides and other chemicals, urging that in addition to research, the provincial government establish a voluntary agricultural chemical management certificate program to educate farmers in chemical use and safety, and that governments require all pesticide company labels and brochures to contain instructions on safety precautions and safety equipment. In the same brief the Women of Unifarm board requested the establishment of a farmer's health branch within the department of health and that the government take some responsibility for educating the medical and health community about pesticide poisoning.

Farm Safety

The 1960s brought the first local activities focused on farm safety. The Fort Saskatchewan local recorded in 1965 that a meeting was devoted to a farm safety discussion. In 1969 members of the Jefferson local participated in a farm hazards and safety school in Banff. This issue would eventually lead to the development of several successful projects, including a series of safety workshops conducted throughout the province in

Kickoff for 1988 Women of Unifarm Farm Safety Hike, May 30. Front, left to right: Louise Christiansen, Margaret Blanchard, Hon. Peter Elzinga, Jacqueline Galloway, Karen Gordon. Second row, left to right: Kate Horner, Jenny Bocock, Ruby Ewaskow, Janet Walter, Judy Pimm, Mary Newton, Joyce Templeton, Mary Wright, Marylou Blakley, Elizabeth Olsen, Sonja Hudson.

1982 and the farm safety hike, which would receive national and international attention in the later 1980s. In 1983 Women of Unifarm received special recognition for helping to reduce the number of children's deaths on Alberta farms. The number of children who died in farm accidents dropped from 16 in 1981 to one in 1983 and farm injuries were also down, based on statistics gathered by Alberta Agriculture's farm safety branch.

The 1984 safety co-ordinator for Women of Unifarm, Louise Christiansen, assisted the Farm Safety Program of Alberta Agriculture to identify the disabled farmers in the province, with an interest in starting a new program focusing on farming innovations. The 1986 convenor, Elsie Seefeldt, wrote as she retired from her position: "Farm safety continues to be an extremely important part of our lives, and Women of Unifarm have an opportunity to really make a mark for ourselves in this area. It is a job that requires time, effort and imagination as well as a real desire on our part to take a leading role in this important area."[18] New legislation on dangerous goods were part of the focus on safety in 1986, as many of the items carried by farmers became classified as dangerous goods and farmers needed to learn the government regulations concerning the labelling and transporting of such goods. In 1987 the Region 11 conference featured a presentation and discussion on all-terrain vehicles. The same year, the new safety coordinator, Jacqueline Galloway, developed an issues study for all locals on farm safety, which led to the launch of the farm safety hike in 1988. The hike grew in popularity in subsequent years. Aided by an illustrated booklet, farm families were encouraged to hike around their own properties, identifying hazards and sites for improved safety measures. As the safety coordinator in 1989, Florence Trautman, pointed out, the participation of the whole family is the key to the hike's success. The farm safety hike has proven to be a successful model program in which other provinces and countries have shown some interest.

Also in 1987 concern for rapidly declining membership figures and few younger women being attracted to the organization prompted a study on how to promote the organization. Several workshops on the same issue, which were assisted by Alberta Agriculture, were offered. In their annual submission to the agriculture and rural affairs caucus committee in 1987, farm women included their resolution concerning the Canadian Rural Transition Program, to allow farmers to receive their training while still residing on the farm, thus relieving some of the family stress associated with having to leave the farm. The brief also requested that the provincial government continue to provide incentives to farmers for planting and maintaining shelter belts, tree lines and blocks of treed land. In their 1989 brief to the caucus committee, Women of Unifarm asked that a rural child care strategy be developed and that future farm employment programs, such as the 1989 Alberta Farm Employment Program, be expanded to include domestic duties so that when farm women hire someone to provide child care, meal preparation, light housekeeping and book work duties essential to the running of the farm could be included. They also requested government support for the establishment of a Centre for Agriculture Medicine, similar to the one founded at the University of Saskatchewan. They closed their brief with a strong belief statement: "Government policy should support the family farm system of food production and make it possible to be viable" because they feel it is the "most efficient and most effective."[19] Farm women of Alberta recognized their industry was changing, and family farms were disappearing. They maintained for many years a belief in farming as a business and as a way of life, and strove to keep their concerns about the survival of agriculture and of rural communities on the public agenda.

National Networking

Farm women of Alberta always looked beyond their provincial borders to exchange ideas and build alliances with other Canadian farm women. They put their belief in the power of organization into action very soon after setting up their provincial organization. Beginning in 1918, farm women across the country began to talk of forming a national association. The concept was centred on the idea of coordination, as they saw the various political organizations of farm women, now active in Alberta, Saskatchewan, Manitoba and Ontario, in need of a national forum for dealing with the "vast amount of national business that is constantly before the organized farm women."[20] The Interprovincial Council of Farm Women was formed at Brandon in January 1919. The Council focused on issues common to all farm women, but its members also had another goal in mind. They wanted to provide a national forum for farm women that was affiliated with the Canadian Council of Agriculture (CCA). In September 1919, they accomplished this goal, becoming the Women's Section of the CCA. Here too women sought "greater oneness with the men"[21] as they worked to bring the farmers' movement into politics, and promote the farmers' interests. The Canadian Council of Agriculture only lasted until 1932, and when it folded, the Women's Section also disbanded. Much of the political power of the CCA lay in its partnership with the Progressives in federal parliament. In 1923 the CCA withdrew its support of the Progressives. In response to that decision, Violet McNaughton, of the Saskatchewan Women Grain Growers and President of the Women's Section of the CCA, resigned, and the women lost their most capable leader.

The Canadian Chamber of Agriculture was founded in November 1935, originally as an organization of dairy and cooperative interests and commodity groups. It reorganized in 1939 as the Canadian Federation of Agriculture (CFA) because all the existing provincial federations of farmers were invited to join. Most of the provincial federations were formed in the period between 1939 and 1949. The CFA is also affiliated with the International Federation of Agricultural Producers. Farm women did not have a national council or section of their own affiliated with the CFA. Although two places were reserved for women on the national council of the CFA, one for a woman from eastern Canada and one from western Canada, they did not have full voting privileges as members of the board of directors until 1992. After years of lobbying by farm women, the constitution was revised to give the two women directors full voting privileges at board meetings in addition to their privilege of voting at the annual general meeting.

Despite the lack of a national organization of farm women, women in farm organizations across Canada maintained an active network between their organizations and an effective lobby of farm women's concerns, particularly in the period between 1980 and 1987. In 1987 the federal government held a national meeting of farm women (the first was held in 1980) that proposed more formal structures to the loose association the women had enjoyed and successfully employed. By 1989 the federal government decided that a national farm women's network should be established whose members would be only the provincial farm women's networks, no farm women's organizations per se could join as individual members. Alberta had established a network of five women's farm organizations in 1986, and Women of Unifarm participated as a founding and key member of the network. After considerable thought, the Alberta network decided not to become an official member of the national network, although it participates in national network meetings as an affiliate and remains in close touch with its members and its activities. (This development is more fully discussed in Chapter Five.) The national network organization formed by the federal government, by the nature of its conception, constitution and mandate, is not a political vehicle for the concerns of Canadian farm women. Farm women have not had a collective political voice on a national level since the Canadian Council on Agriculture was disbanded.

Internal Politics

The internal politics between organized farm men and women in Alberta proved to be the most difficult to understand and resolve as each generation came into the organization with new vision and new energy. Despite the ongoing rhetoric about working side by side with the men, there were issues important to women of the organization over which men exerted their controlling interest as a "parent organization," as Vera Rude referred to Unifarm in 1981. This control was perhaps most strongly felt throughout the years on the issues of women's property rights and women's reproductive health rights. Certainly the rule remained that any lobbying women did with provincial or federal governments must be on policy positions with which the men agreed, though in Unifarm days this came to be understood as mostly economic policy resolutions and recommendations.

The long campaign women waged on the issue of equal property rights demonstrates the position of women both in agriculture and in the organization. The Dower Law campaign was long and hard fought, as each revised piece of legislation introduced after the first law was enacted in 1917 proved unsatisfactory in some ways. This campaign led to a discussion, first initiated in 1919, of the idea of community property, a concept successfully adopted into law in other jurisdictions in the United States and Europe. After the progressive community property legislation was introduced by Irene Parlby in 1921, studied by a committee of the legislature and the whole idea shelved in 1929 by the Farmers' government, the women's organization repeatedly tried to improve the property allowance to which farm women were legally entitled. They introduced a resolution to amend provincial law to allow women to own one third of all property acquired after marriage in 1948, but had to reintroduce the resolution every year from 1952 to 1960 before any action was taken and even then, the results were unsatisfactory. The property laws for women were reviewed again at the 1961 to 1964 conventions, again without achieving anything near equality in property rights. Hazel Braithwaite observed "Quite an improvement has been gained over the years, but it's still not good enough. I still feel women have a long way to go to equal status. . . . Every farm woman should be a member of the farm union. And property laws are the first thing they should be working on."[22]

As the Women of Unifarm organization became involved in 1973 in an appeal to the Supreme Court for some fairness in the division of property for Irene Murdoch upon the dissolution of her marriage, they formulated a new policy on matrimonial property. The case garnered national attention, required substantial financial and time commitments from the organization, and galvanized farm women's determination and efforts to at last have this legal discrimination addressed. Irene Murdoch had requested action by the Supreme Court of Alberta to declare her a partner in her family's farm. This entailed receiving recognition for her financial contributions to the farm, which was legally owned by her husband, as well as her 25 years of unpaid labor on the farm, including five months of the year when she was sole operator while her husband was away from the farm employed in wage labor. She had also requested a legal separation. When her request for partnership failed she appealed to the Supreme Court of Canada. Betty Pedersen was President of Women of Unifarm at the time and it was she who instigated the hiring of another lawyer to assist Mrs. Murdoch, as well as awareness and fund-raising campaigns. When Mrs. Murdoch's case was unsuccessful in the Supreme Court of Canada ruling of 1974, farm and urban women were shocked at the lack of recognition of Mrs. Murdoch's contributions to the family business. The issue of matrimonial property rights was once again put on the agendas of every women's group in the country.

The Women of Unifarm briefs to the provincial government throughout the 1970s addressed the issue of matrimonial property, repeatedly asking for an even split between married partners upon dissolution of the marriage of all property acquired after the marriage. Gifts and inheritances would be exempt from sharing, as well as assets existing at the time of marriage. In 1976 the women also had to address the Report of the Alberta Institute of Law Research and Reform that stated that although in principle husband and wife should share the economic gains they made, that this should not apply automatically to couples already married and living in the province. Women of Unifarm urged that any legislative changes that were to be made include all existing marriages. Finally in 1978, after the Murdoch case created a new public awareness and concern, legislation was passed in Alberta that reflected the policy for which farm women had lobbied for decades; all matrimonial property could be divided evenly between spouses upon dissolution of the marriage. The legislation left the decision about what constituted matrimonial property to the discretion of the judge and many judges did not and still do not acknowledge women's unpaid labor. The best protection for farm women is a legal partnership in the enterprise. Still, even in the early 1980s less than half of Alberta farm women had arranged for a legal partnership with their husbands in family farms, despite the equal contributions they made to the development of the family businesses.

Beyond the concrete issue of property rights, farm women continued to identify the secondary status they held in farming. In a 1977 report on farm stress, they ranked family stress, which centred on "lack of recognition for the female role in the farm family" as the second major stressor, after finances.[23] Farm women also had to contend with the attitudes of their spouses towards their organizational interests. For some women, there was unqualified support, as husbands valued their wives' contributions to the farmers' cause, as well as the positive effects organizational involvement seemed to have on women's outlooks and well-being. For other farm women, their involvement in the farm women's organization was a source of tension and even hostility between husband and wife, as the work took them away from their work on the farms and beyond men's influence. Women were also more frequently than the men requested to curtail outside activities, particularly organizational participation, when the demands of the farm required more attention or because men felt threatened.

With so many barriers to overcome, it is remarkable that so many generations of women accomplished so much on behalf of the farming industry in Alberta. Farm women took their experiences in the local farm organizations to other groups, including commodity organizations, co-operatives, environmental organizations and task forces. Betty Pedersen listed in her memoirs, for example, nine provincial committees and task forces and one national council on which she served while she was both president of Women of Unifarm and a director of Unifarm. They took their interest in farming as a viable way of life to national and international forums, including the Associated Country Women of the World, the Canadian Federation of Agriculture and the League of Nations (see Chapter Five).

The political activity of farm women has roots in the homestead experiences of women who were trying to survive harsh conditions and build a decent life for their children. From these roots grew a strong organization that for many years was poised to tackle any issue that affected the quality of life in farm communities. Women who learned their activism in this organization spread their influence to other organizations. Boards, commissions and committees throughout Alberta were blessed with the insight and commitment of farm women as they brought their perspectives and knowledge to the proceedings. Farm women have branched out in recent decades, selecting causes and committees and organizations where they can be most effective in addressing the issues of most importance to their own families and their own communities. They have also branched out by forming liaisons and networks with other organizations. Regardless of the venue they chose, farm women remained aware and involved in the political decision making that shaped their lives. Indeed, they have rightly earned a reputation for their activism. The often repeated statement of Helen McCorquodale, while she was editor of the *High River Times*, aptly described this reputation:

> *There will be no difficulty in recognizing the Alberta women in Heaven. With pencils and notebooks they will be gathered in little groups beside the river of life, putting finishing touches on resolution B2894, urging that more rural children shall be taken into the heavenly choirs.*[24]

Farm Women's Creed

That true cooperation practised in the home and community life makes for the highest type of character building.

That the Master of Life governs by unfailing laws which are beneficent and stabilizing.

That when we learn to know and adapt ourselves to those laws we find contentment and happiness.

That to live one day at a time is the only way.

That to judge others is not our task.

That to free our minds from weedy thoughts that crop up, we must plant and cultivate the flowers of loveliness in their place.

That love is the highest and most potent of all laws.

(from the 1948 United Farm Women of Alberta Program)

Chapter Three

Creating Community

The founding members of farm women's locals brought to their gatherings their own needs and experiences as women, mothers and farmers. Even before the provincial organization was born, farm women met together to socialize and to organize. They found as they connected with each other at events such as childbirth, picnics, teas and eventually women's meetings, that they had common concerns and a shared vision of community improvement. They identified early the needs of children, mothers and the less fortunate in their farm districts. Most pressing were the needs for appropriate medical care for expectant mothers and for family illnesses and accidents. Most mothers worried about the opportunities for their children to be educated, and to have some of the advantages, like libraries and social activities, that the mothers themselves had enjoyed as young people. It did not take them long to get to work. And when the opportunity arose to link their efforts to those of others, to speak with increasing strength and growing numbers to those who controlled the province's treasury, farm women seized it with enthusiasm and energy.

The first three committees struck by the new United Farm Women of Alberta were the mainstay of its work for many years: health, education and social welfare. Farm women took responsibility for these issues because they were keenly interested in them and they were aware that they were not priorities for the men. They became willing students of social policy and effective organizing. As Betty Pedersen described it, "Well, the women have always been given certain areas to handle . . . the health and the education and the social welfare. The men pretty well left it to them and said, 'that's women's work, now you do it'." Organized farm women's interest and participation in the development and improvement of social, health and education services for rural Albertans endured through all the years of the organization.

Farm women are justifiably proud of their record in these areas. As Louise Johnston said, "Well I think our influence on health and education and social welfare has been very positive right from the early years of the organization." The influence and dedication of farm women can be seen in the number of laws that were changed or introduced as a result of their lobbying. Betty Pedersen observed, "As far as changing the laws are concerned, I think the women have done a great deal, they haven't done

an outstanding job but I think they have done better than the men have."
In addition to social reform evident in new laws, farm women joined
every committee and association table that concerned itself with the
welfare of Alberta citizens, including special groups, like the Indian/Es-
kimo Association and the Hutterite Committee. Louise Johnston explained, "It [the farm women's organization] was the social organization
that was able to take a hold of anything that was coming up . . . whatever
a local was interested in, it was vital in that particular community, so that
was the organization that picked it up." Miriam Galloway also saw the
organization as "an opportunity to solve problems particular to rural
people in education and health and social services. And it has proved
that we can change things."

Sometimes the leadership of the organization, the interests of the
president and other executive members, would influence the choices
locals or the board made in directing their precious volunteer hours and
funds to one cause or another. In many cases, the personal experiences
and the professional training shaped the interests or favorite causes of
the president. Irene Parlby, the first president of the organization, was
particularly interested in maternity care and mental health. Susan Gunn
was also very concerned about the adequacy of mental health services in
the province. Mrs. Price, a president in the 1930s, stressed birth control
measures for mentally ill persons. Louise Johnston, as a teacher and a
socialist, was interested in education and the cooperative movement. She
was a president in the 1960s. Betty Pedersen, a strong feminist and a
president in the 1970s, was interested in women's property rights and
legal status, planned parenthood and sex education in the schools. Leda
Jensen followed Betty Pederson and carried on her interest in planned
parenthood and also in leadership training. Ivy Taylor, a retired school
teacher like Louise Johnston and Betty Pedersen, brought her passion for
education and her interest in surface rights to her presidency. Margaret
Blanchard, president for two years in the 1980s, was very interested in
mental health.

Health

Farm women who speak of the accomplishments of the organization
recall most frequently the work they performed in making rural communities good places to live, where quality care for medical and other needs
could be readily accessed. They organized local resources and lobbied
municipal and provincial officials to pay attention to and provide funding for needed services in public health, acute medical care, reproductive
health care, mental health and services for disabled and elderly persons.
Farm women were relentless in their pursuit of quality health care for
rural citizens. Projects and issues varied widely in size and impact.

The first concern of the United Farm Women was appropriate care for
expectant mothers and their babies. Correspondence between Irene

Parlby, the first president of UFWA, and her Saskatchewan counterpart, Violet McNaughton, shows that the need for trained birth attendants was acute and topped the list of demands farm women placed before governments. Statistics showing high maternal and infant mortality rates confirmed the many stories of tragedy they heard from their own families and neighbors that were common across the province. Farm women used these statistics to support their demands to government. As Winnifred Ross said, "The interest in improving health care for rural people was more from the death of babies as it was higher than third world countries at the time. There were no services that we had to rely on. We used midwives but they had no training, they were just volunteers really."[1] It was difficult to find solutions to the maternity care needs of women in remote or isolated areas and for women who could not afford the expensive care of doctors. The municipal hospital scheme, the public health nurse and district nurse programs, the travelling clinics and the establishment of government health insurance were all measures instigated by farm women who saw these as essential services to farm families. These approaches, however, were proposed only after the midwife scheme of 1916, presented by organized farm women of Saskatchewan and Alberta to a national task force on maternity care (sponsored by the Canadian National Association of Trained Nurses) was rejected by Canadian nurses and doctors.[2] The hospital programs never met the needs of mothers who lived too far away to use them. The major barrier to most families who required medical care of any kind was the high cost. Individually, and then later collectively through their organization, farm women insisted that health care had to be organized by government, and had to be affordable. In 1919, as the Alberta Public Health Department was being formed and the first obstetrical nurses for remote districts were being trained, Irene Parlby proposed the first universal government medical insurance scheme to the federal government on behalf of the United Farm Women of Alberta. Her idea did not carry much weight with the government, although some medical services to families became subsidized by the provincial government, through minimum payment schemes, and by hiring some salaried doctors and nurses on the government payroll, particularly in remote areas.

For decades, farm women continued to lobby for a government-sponsored health care system. Louise Johnston remembered that most of her years as president of Farm Women's Union in the 1960s were spent lobbying the government on health insurance. But before government health care insurance was implemented, families who were members of the organization could participate in a group medical insurance plan (MSI) adopted by the Farmers' Union of Alberta. It was primarily farm women in each district who acted as secretary-treasurers for the health insurance plan, gathering people's payments, helping them with claims, and in some cases, sponsoring a family in financial distress out of their own pockets so that its coverage would continue. As some farm women who performed this task pointed out, it was a huge responsibility added to an already full work day.

In 1949 Jean Rose reported that in the Ponoka area the health care focus was on the establishment of health units and the immunization program. In the same year, the executive of the FWUA presented a brief to the Alberta Health Survey Committee (see Appendix B). As Mrs. Vera Lowe, president of FWUA, described it to convention delegates, "the brief was prefaced by reiterating as our opinion [that] nothing less than National Health Insurance under the Dominion Government, in co-operation with the provinces, can provide adequate public health and medical care service."[3] In 1961 the annual convention passed a resolution requesting that a national health insurance plan under federal government sponsorship and control be put in place "to give full medical, surgical, optical, dental and burial facilities at a premium that the lowest income group can reasonably afford."[4] At the same meeting, Mrs. Trew of the Saskatchewan Farmers' Union Women's Section spoke about the Saskatchewan Health Plan, and delegates asked Mrs. Trew "many questions." After medical insurance was adopted as a government program, farm women continued to press for a dental health care plan for children. Baby clinics were another project of the early days, and these were gradually taken over by the public health nurses when they became established in a district.

Betty Pedersen mentioned that farm women lobbied persistently to have senior citizens' lodges established in Alberta, and as she described it, "we got them, but one thing we didn't ask for, we were too stupid to ask for, was private rooms. Instead of that we got lodges that were all double rooms. Now that has changed. After my time [as president, 1970-1975] that changed because they kept on lobbying that there was not decency and privacy for an older person to share a room with another person and many of the lodges are now changed so that you do have a private room."

Emma Innocent recalled the work of women in her local, at Roseland, to get the "Brompton cocktail," a drug mixture administered by doctors to terminally ill patients to ease pain, legalized in Canada. They canvassed doctors in their area, and with doctors' support, pressed the provincial government to promote its legalization. The Canadian government legalized its use shortly afterwards. The Roseland local also surveyed all the hospitals and auxiliary homes in their area, and were instrumental in the facilities' receipt of more grants and improved nursing service in auxiliary hospitals.

In many communities, farm women worked diligently over the decades to ensure hospitals and other care facilities were properly equipped. The Griffin Creek local furnished a room in the district hospital, as did the Berrywater local; the Westlock local donated a lung machine to their hospital and also contributed to the purchase of a home dialysis machine; many locals sewed or raised money for Red Cross; and

7th Annual Convention, Farm Women's Union of Alberta, Edmonton, Dec. 6-9, 1955. (photo by Wells Studios of Edmonton)

raising money for cancer research and services was common to all locals at some point in their histories. Hazel Andersen pointed out that her local, at Freedom, worked for years with the hospital auxiliary. Many seniors' lodges and hospitals would not have continued to function without the extra help and donations provided over the years by some farm women's groups. The seniors themselves benefited directly from the social functions many farm women organized regularly, including plays, parties, musicals and concerts. Health care organizations and societies thrived due to the fundraising work of farm women. The Heart Fund, the Multiple Sclerosis Society, the Red Cross, the Tuberculosis Association and the Cancer Society were all recipients of funds raised by farm women. In 1958, the board of the FWUA sent a questionnaire to all the locals, asking about their charitable donations. Although less than half of the locals responded (73 out of 161), the results of the survey were impressive. Twenty-six different organizations and services were listed as recipients of monies from the farm women's locals, for an estimated total donation of over $17 000. Most of the recipient groups were health-related ones, followed by social welfare agencies, then scholarships and young people's work. Cancer and mental health associations were the two most popular recipients of local donations.

Wartime brought an increased effort by farm women throughout the province to supply the Red Cross with knitted garments, sewn items and food parcels. The Westlock local members helped to equip a Red Cross mobile unit to be sent overseas during the Second World War. The Sunnyvale local sent the huge sum of $120.00 to the Ramsgate Convalescent Home during World War I, $100.00 of which was to equip two beds and $20 to go to the work of the Red Cross. Mabel Barker's sharpest memories, at the age of 100, were of the war work her local did in support of the Red Cross:

> *During the war, the two wars we had, we did an awful lot of work for Red Cross. I used to go to the depot in the old city hall and pick up a whole car load of flannelette and old Mrs. Watson and old Mrs. Hobson and Mrs. Ellis, there was three of them, they would cut out the pajamas and then sew it. And then we did a lot of knitting. At that time you had to do the mitts. But you had to cast off here at the end of the fingers so that they could pull their fingers outside of their mitts on account of the war. And the women didn't like doing the mittens so I used to knit a lot of the mittens but they'd sew the pajamas.*

Mental health services were a concern of farm women too. It was in response to the demands of farm women that the government constructed the first mental health facility at Ponoka. Concerns in the 1920s reflected a belief in the spread of mental illness from one generation to the next, and mental health discussions were centred on sterilization of mental patients. In 1949 the FWUA concentrated on the study of mental health "from the preventive side," and Dr. Hincks, director of the National Committee on Mental Hygiene, spoke to the FWUA convention that year to "outline the work and the programme."[5] Many locals over the years not only raised money for what became the Canadian Mental Health Association, but also had active members participate on mental heath boards and committees. In addition some locals had a direct relationship with the Alberta School Hospital and the hospital in Ponoka, assisting particularly at Christmas with gifts. Farm women lobbied for many years to have mental health services offered in conjunction with the services of the rural health units, and that mental health patients be given "hospital, medical and treatment free of charge."[6] In 1958 they also requested that the federal government provide more financial aid for mental health research.

In 1959 the FWUA petitioned the government to extend full medical coverage to cancer patients, for whom only seven days in hospital were covered under medical insurance plans. The 1977 convention requested that a Department of Gerontology be established at one of the Alberta medical faculties, noting that Alberta was "27 years behind the times in its care of the elderly."[7]

Reproductive health would prove to be the most contentious issue the women dealt with over the years. Issues such as birth control, sexual sterilization and abortion were particularly difficult. No consensus on these issues was ever achieved, and their divisive character was further reinforced by the men's disapproval of the women's positions on some of these issues. The birth control debate emerged quite early in the organization's history, in 1919 as the First World War ended and the concern for the safety of mothers was heightened by awareness of maternal death rates and the still primitive conditions in which many women faced pregnancy and childbirth. In 1922 the convention delegates passed a resolution in favor of birth control information being made available to married women. Another resolution in the early 1920s was passed requesting that contraceptives be made legal and available. The resolution passed only after heated discussion for two and a half hours. Those against the resolution stated that they thought the use of contraceptives was "immoral, irreligious and shameful."[8] Those in favor of the resolution thought it was a blessing, as it enabled mothers to space the births of their children and thus the health of the mother would be safeguarded instead of a new baby every year regardless of the consideration of mother's health and vitality. At one point in the debate a delegate shouted:

> *This would make marriage an emotional playground wherein women would romp through the years like a courtesan in carnal enjoyment — far from the biblical injunction that in sorrow and pain women should bring forth children. In explanation of the virginal mother Eve it was destined that women were ordained to suffer and submit.*[9]

Perhaps the most controversial public debate with which the women became involved was the sexual sterilization of mentally deficient persons. As with many issues, Alberta farm women sought in-depth knowledge about the issue, and studied the latest scientific research and expert advice on the matter. This is exemplified by an address by Irene Parlby to the 1924 UFWA convention in which she stated, "in fact investigations in every country show that mental defectives are reproducing their kind at an alarming rate and that society for its own protection must take more adequate steps to control the situation."[10] In her address Parlby was careful to review the latest knowledge on the subject and to recommend thoughtful consideration of the problem. She did not support sexual sterilization as a solution to a growing population of mentally ill persons. In fact, despite the fact that the United Farm Women Convention of 1927 passed a resolution put forward by their Convenor of Health and Child Welfare, Mrs. J. W. Field, to recommend to the government that they introduce legislation to deal, in some degree, with the problem of prevention of the procreation of the mentally unfit, not all farm women were comfortable with this position. Mrs. Field provided in her report to the convention a summary of the legislative efforts and their consequences in other jurisdictions, making it clear that laws permitting sterilization were not introduced without creating controversy or being legally challenged. Farm women were aware of the concerns about a growing population of mental patients dependent on government care. They had been active participants in ensuring appropriate care was available throughout the early years, and had been invited by the Minister of Health to participate on an institutional visiting committee, which was established to monitor the care and standards of every mental health institution in the province. Their position on the sterilization of mentally-deficient persons came from an intimate knowledge of the care system and a faith in the mental health experts of the day.

In 1928 the Sexual Sterilization Act was passed for the sterilization of mental patients, under the authority of the Eugenics Board. It was supported by the UFWA, based on the popular belief that mental illness was hereditary. An important feature of the legislation, which was later amended, was the requirement to obtain patient consent, or in the case of a patient's mental inability to provide it, to obtain permission from the family. In 1931 Mrs. Price, first vice-president of the United Farm Women of Alberta, urged convention delegates to undertake an "extensive study and investigation of birth control and its relation to social and economic conditions."[11] The issue of birth control had never been singular in focus for farm women. The presentations to government always included the availability of birth control information and procedures to married women through "family limitation clinics" and sex education in the schools, along with recommendations about sterilization being available to those who were mentally unfit to be parents. There was an overriding philosophy that all babies should be wanted babies, brought into circumstances that would be healthy for babies and mothers alike. The additional impetus behind the push for sterilization of mentally-deficient persons was the impact that the care of mentally-dependent persons was having on the provincial health care budget – 80% of the health department budget was spent on mental health institutions and services, which also constituted 60% of the total provincial budget. In referring to birth control only in terms of sterilization of the "mental deficient," Mrs. Price asked the 1931 convention delegates "Are we losing courage, that we have let this extremely important matter, once discussed, drop?"[11] The views on the sterilization of mental patients as a responsible social measure to curb, either through controlling genetics or environment, a growing mentally-disabled population, continued into the '30s, as a 1933 health bulletin identified.

Farm women also requested that a medical examination before marriage be introduced by the government, sending a resolution to government on this issue in 1932. Farm women saw this as another strategy to prevent the spread of hereditary diseases, including mental illness. The Honourable Mr. Lymburn responded on behalf of the government, stating that he "did not think public opinion was quite ready for that yet."[12]

The 1930s also brought a renewed call by farm women for government sponsored "family limitation" clinics for married women. Resolutions to this effect were passed and reaffirmed by the UFWA and endorsed by the UFA board in 1933, 1934, 1935, 1936. The subject of birth control was a major focus of the 1933 convention, with the health convener, Mrs. Banner, devoting most of her remarks to it and the presentation of an address on the subject to the convention by Dr. Folinsbee Newell. A reply from the government on the issue simply promised "careful consideration," but no action was taken. The issue was a constant concern for the organization, which for decades joined many other women's groups to press the federal government for the legalization of birth control. Finally in 1969 the federal government legalized the use of birth control in Canada. In 1970 Betty Pedersen, the president of Women of Unifarm, was a founding member of the Alberta Birth Control Association, and later forged a strong link between Planned Parenthood and the farm organization, despite internal disagreement between farm women about this affiliation. The organization maintained an affiliation with the Alberta Birth Control Association, later renamed the Alberta Family Planning Association, and with Planned Parenthood, despite the withdrawal of many long-time members upset about the abortion referral services of Planned Parenthood. Heated and angry debates over birth control in the early years were replaced by new ones about abortion in the 1970s and 1980s. Although much of the disagreement between parties in both cases was based on religious beliefs, the findings of the Badgely Report and the participation of Leda Jensen, president of Women of Unifarm (1976-1979), on the board of Planned Parenthood, were cited as reasons for some farm women's withdrawal from the organization, and for demands for Women of Unifarm to break its affiliation with Planned Parenthood. In the 1970s there was support for the decriminalization of abortion. In the 1980s some Women of Unifarm members maintained the policy position that a woman's right to an abortion should be determined solely on the basis that the pregnancy endangered her own life. In 1982 Region 7 conference passed a resolution protesting the establishment of private abortion clinics and asking that "laws be established to ensure that such clinics are illegal."[13] However, the organization continued its affiliation with the Alberta Family Planning Association and maintained a representative on the Planned Parenthood Alberta board.

Venereal disease was also a cause which farm women took up with candor and compassion. In 1919, they began a long history of campaigns on this issue by requesting that the government make the disease reportable and under the health regulations. They urged "that those suffering from syphilis should have three year treatment and those suffering with gonorrhoea a one-year treatment," stipulating that these treatments would be government funded.[14] In 1935 they launched a public education program on venereal disease and petitioned the government to offer some form of sex education through the Department of Public Health.

In more recent years, after a fully-funded medical system was in place, the needs of particular groups of patients were the focus of requests to the government. For example, in 1985 the Eastburg local wrote to Dr. Russell, Minister of Health, requesting that daycare centres be developed for brain injury recovery patients.

Education

The educational opportunities provided in rural areas were as important as health care to farm women. For many years, it was an accomplishment to have a school operating in a rural community most of the year. A bulletin on education by Winnifred Ross in 1932 described the struggle to keep rural schools operating. "The problem of financing the operation of schools which became acute in the beginning of 1931 has continued without relief, and constitutes the greatest problem [with] which the Government, school officials generally and teachers have to deal."[15] Later, as rural communities prospered and the schools were well established, farm women focused on the concern for a quality of education comparable to that provided to children in major urban centres.

A second major educational goal was to maintain agricultural education at various levels: at the agricultural schools for post-secondary training, and in the elementary and high school system so that children would have a basic knowledge, regardless of their location and experience, of the importance of agriculture to the life of the province. Hazel Andersen remembered that the organization "really pushed for agriculture to be taught in the school curriculum." In 1961 the FWUA pressured government to set up an irrigation agricultural school in southern Alberta. This lobbying led to the development of a course in irrigation at Lethbridge University. The organization had representatives participate on various councils and boards of the Rural Education and Development Association (REDA), the Agricultural Education and Rural Extension Advisory Council and the Agricultural Awareness Foundation board of directors.

Continuing and adult education were also concerns, and this included the provision of services like libraries and noncredit courses. Libraries were a focus of many of the locals' interest and energies. One of Irene Parlby's first projects in her community of Alix was to solicit books from abroad to create a community library. Other locals participated in travelling library projects. Margaret Richardson of the Berrywater local, whose mother, Mrs. Oldfield, was the first president of the local in 1919, wrote, "our local contributed much to the organization of travelling libraries, and I remember when one of the big boxes of books arrived at our house."[16] When town libraries were established, many locals regularly donated funds and books to them, sometimes in memory of deceased members. At the first annual meeting of the Sunnyvale local on December 9, 1915, the members decided to develop a library. The High Prairie local is proud of their work in building the museum and the first library. They still contribute to the library, and also run the museum, their biggest responsibility. Farm women lobbied for the establishment of the regional library system and the grant program to support them. The organization was also responsible for the introduction of the provincial school book rental plan, that allowed students to pay a nominal fee for the use of textbooks, instead of incurring the expense of purchasing them.

The Roseland local was instrumental in starting a trade school for slow learners in Camrose. After investigating the need among families of the area, a resolution developed by the local was passed at the women's spring conference at Forestburg, taken to the men's conference, and then, as Emma Innocent described it, "it just seemed it went like wild fire" after she "did the talking in Edmonton." The farm women of that district are justifiably proud of the Centra-Cam School in Camrose, for it would not have been built in the community without their commitment to it.

The farm women's organization was part of the Alberta Education Council for many years, and later the Alberta Association for Continuing Education and The Canadian Association for Adult Education. For a number of years Louise Johnston represented the farm women's organization at the Alberta Education Council, after serving as education convener of the FWUA. She eventually became the Council's president. Other FWUA members served in advisory capacities on various educational committees. These included the advisory committee that set up the dental assistant program at the University of Alberta on which Florence Hallum served, and the advisory committee on agricultural education. Six farm women who were board members also served on the Senate of the University of Alberta. Members of the organization were also very involved in surveys on education and contributed to the Cameron Commission's study of education. When the Commission's Report was released in 1960, the president of the FWUA, Hazel Braithwaite, outlined a program of study of the report for board members, and suggested to all the locals that they undertake a study of the report as well. From their study of the report, the FWUA formulated many resolutions, debated a number of issues and supported many of the report's recommendations. All of these activities gave the provincial government useful feedback about rural people's views and concerns about the education system.

In 1961 the FWUA Board requested that the Department of Education "compile a set of examination standards for each grade in Arithmetic, Language, Reading and Spelling" and "that these be changed each year and be administered at Christmas or Easter."[17] The board also put forward a resolution that the teaching of another language begin in the lower grades in school. An important resolution was also introduced at the convention of 1961, requesting that "provision be made to educate a child at public expense thoroughly from kindergarten through academic, vocational, trade school or technical institute, suited to the individual."[18]

In 1963 the FWUA Board asked the Department of Education to provide a special training course for teachers who were willing to teach retarded children. Another initiative that farm women were "very involved in" was "getting family life education programs in to the schools." Betty Pedersen explained, "we really lobbied on that one and we got it." Farm women also lobbied consistently for the semester system in high schools. "It's hard to change the law," Pedersen stated, "but we have had an influence in changing laws and that was our objective."

The members of the Grandmeadow local felt that they "contributed a lot to children's education, gave them more opportunities." They mentioned in particular the young people's week at the university that was sponsored by the organization, and which was offered from 1919 to the late 1960s. In their words, "It gave those young people an idea of what was out there."[19] Farm women were also concerned about the plight of gifted children in the school system, and for many years they pushed for both recognition of the rights of gifted children and for programs to be developed for them. In 1977 the organization lobbied the education minister, Julian Koziak, for more action to deal with the problem of learning disabled students. In-service teacher training, more training for future teachers and more funds for programs and research were part of the request. In 1983 the board of directors sent a brief to the Task Force Review of High School Student Evaluation.

The organization sponsored a number of scholarships for young people to further their education at the post-secondary level, either in agriculture or in other programs of study. The Irene Parlby Scholarship was the first, developed in 1963 for female students at the three agricultural colleges. The Inga Marr Memorial Scholarship was created after her untimely death in 1975. It is awarded to a young person of rural background who has participated in 4-H programs and is enrolled in post-secondary education. They sponsored the circulation of a scholarships and bursaries list to all grade nine and grade 12 students. The organization also donated to the Lady Aberdeen Scholarship Fund of the Associated Country Women of the World (ACWW).

At the individual school level, farm women's locals pushed for programs like music and band. The Freedom local was one of these, formulating resolutions and meeting with county councillors to convince them that music education was important. Also at the local level, some farm women's locals identified particular post-secondary students who could use a little financial aid, to buy textbooks or pay tuition, and funds were provided from the locals' accounts.

In a broader community-based way, farm women were responsible for educating Alberta citizens about cooperation and cooperatives and how they work. Margaret House felt strongly about the important role farm women played in setting up cooperatives and making them successful. From the beginning, farm women participated in cooperative buying and marketing strategies. Some of these were formalized into the Co-op stores, still run by community boards. Some farm women were members of the Co-operative Women's Guild for many years. A large international organization with 20 000 000 members in 29 countries in 1961, the Guild was organized to teach cooperative philosophy. At the 1961 FWUA convention, a resolution was passed urging the government to introduce more information into the high school curriculum on the philosophy and history of cooperation, and that students be taught this information when they study business methods. Further, the resolution suggested that methods of teaching about cooperatives be included in teacher training courses.

Social Welfare

The social welfare activities of farm women changed both in character and size, expanding over the years to include more issues and challenges. The focus in the early years was on the immediate concerns of neighbors and residents of an area who were dealing with problems like spousal or child abuse, destitution because of fire or failed crops, alcoholism or desertion of a family by a father or mother. Farm women were always ready to help with food, clothing and quilts, contacting and persuading appropriate authorities, or by raising money to remedy the situation. The McCafferty local, for example, discussed at one of their first meetings the plight of a young boy who was being beaten by the uncle with whom he was living. The women decided to take action, and in consultation with the men of the organization, offered the boy temporary shelter and then found other relatives to look after him. Many locals always had a quilt and other household sewing ready in case someone was burned out or needy in their communities.

Fall 1952 – Forty-five farmers gathered to harvest the crop of the late Jack Hodgson, suddenly dead of a heart attack. Chestermere Farm Women's Union of Alberta served lunch and a chicken dinner to the men. (photo courtesy Dorothy Clayton)

Other projects that directly affected the welfare of the community were restrooms in town for farm families, lodges for the elderly, and library and recreation facilities. The Brooks restroom is an example of farm women's community spirit at work. As one Brooks member explained it:

> At that time [1930s] everybody had probably one means of getting to town. And when the farmer came to town then the wife came with him and she had no place to go to wait until he did all his business and so on. Often they had babies and children to look after and that's why we felt there was a great need for this. And we did quite a lot of the work on it, . . . the painting and the papering and all the different things that we had to do. And collecting money for the restroom, I can remember canvassing the people in Brooks . . . Well it was a building and it had a big room. At the back there [were] quarters for the caretaker and then downstairs underneath we had the first library . . . there were a great many uses made of this restroom . . . It had a crib for babies and Mrs. Robinson [caretaker] would take all of the children. Yes and they had card parties in the restroom . . . there [were] people changing for weddings and things like that . . . well they did their washing in the wash basins too . . . and the working ladies would take their lunch and have their lunch there.

The Brooks restroom operated until the early 1960s when it was torn down by order of town council in the name of "progress." It clearly provided a number of important services to women and children who came to town, and to town citizens as well. It was a focal point for many of the farm women's activities, and a project of which they have always been proud. Through the request of the local at de Winton, members of the Calgary local, with the cooperation of the Calgary Women's Institute, the members approached the Calgary City Council and the Board of Trade about a restroom in Calgary. Both agreed to contribute equally to the establishment of a rest centre for farm families near Calgary City Hall in 1921. Other communities opened restrooms for farm women and their children, often through the efforts of the farm women's locals. The Berrywater local worked for many years, into the 1990s, to keep the restroom in Vulcan in operation.

Temperance was identified in the beginning years as an issue of concern to farm women. Many farm women campaigned vigorously on behalf of prohibition, which became effective in Alberta in 1915. With the repeal of the prohibition legislation in 1923, farm women focused their efforts on the control of sales and consumption of alcohol, with concerns for the well-being of women and children. At the annual convention in 1935, the UFWA president, Mrs. Price, summarized their position in the following way:

> *Year after year, the UFWA at the Annual Convention had gone on record in favour of temperance and education in temperance. Last year our resolution read in part as follows "That we do earnestly request that the Alberta Government do not permit any extension of hours in the Province of Alberta, for the selling of beer". To many of us the disappointment was exceedingly keen when through amendments to the Liquor Control Act, not only was the extension of hours granted, but bottled beer could be sold for consumption elsewhere than on the premises. I know this is a controversial subject, but repeatedly this year I have been asked if we women cannot do something to have this amendment rescinded. The boys and girls are precious to their Dads and Mothers and it is a fallacious argument that they make better men and women by overcoming the temptations we place in their path. No farm women would dream of planting seeds in her flower or vegetable gardens and then sowing sow thistle or any other kind of thistle or weed seed with the expectation that the flower garden would be a thing of beauty, or the vegetables of better quality for so doing. Common sense would tell her otherwise, but we expect the best from our young people, and we are not hesitant in our criticism of them, no matter how careless we are of the social conditions under which they must live and for which we, and not they, are solely responsible.*[20]

Some of the social welfare issues identified in later years were dealt with in a direct fashion, more through concern for the larger community than for specific individuals: day care, hail insurance, rural telephones, pornography, drug and alcohol abuse, child welfare, family violence, adoption policies, safe driving, the family court system, care of the elderly, care of the cemeteries and environmental concerns. The organization always viewed community welfare in the broadest sense, and there was nothing too trivial to try to fix. One member of the Sunniebend local calls these local improvements, such as the road signs they lobbied for to prevent people from driving into a creek, "real issues," because they were things that directly concerned their community. Ruth Wilson of Fairview recalled the work her local did to maintain and improve the cemetery. The Fairview women also planted trees in their district. Hazel Andersen recalled that it was the farm women who started the rural telephones in her area. "We were the big push behind that and we just hounded the men and the government and we're quite proud of that." In the mid-1980s farm women lobbied to have private telephone lines installed to replace the party-line system.

One local near Lacombe opposed the sale of liquor in grocery stores, and lobbied to have the drinking age raised. They also opposed the sale of pornography and Sunday store openings. The 1986 spring conference of Region 1 Women of Unifarm in Eaglesham focused on family violence, suicide prevention and family communication. Professionals involved with each issue in the community offered information and advice.

At the board level, many ideas for reform and social improvement found their way into letters and briefs addressed to all levels of government. The women had long been interested in prison reform, developing resolutions on the issue in 1933, writing a letter of appreciation to Agnes McPhail in 1935 for her work on prison reform, and supporting the work of the John Howard and Elizabeth Fry societies. In 1958 the FWUA Board wrote to the federal government, petitioning for reforms at the penitentiary for women in Kingston, and for the assistance of the federal government in reforming provincial jails for women. Specifically they asked for greater use of probation and parole, segregation of first-time offenders into home-like surroundings, rehabilitation programs that included vocational training, release programs that helped women find a job, a place to live and adjust to life in the community, and a complete training program for all personnel who would be involved in implementing these new approaches. In this proposal, as in many others put forward by the organized farm women, there was evidence that the women had conducted some research into conditions and programs in other countries, and also that they dealt with the issue of women prisoners in a humanitarian and practical way, with the women's best interests at heart. These qualities characterized the approach of the farm women's group to social problems for many years: they were willing to work at becoming knowledgeable about the issue and potential solutions. As a result their reforms always reflected a caring and practical approach to each situation. They also recognized that information or training would be required for those working in the social services if changes were to be effective and lasting.

Another example of the concern for training arose in 1961, when the FWUA asked the provincial government to provide courses in social welfare and mental health, and to allow more bursaries and scholarships for training social workers. In the 1960s, after years of individual resolutions about the situation of Alberta Indians (particularly the young people living on reserves), the organization adopted an Indian Policy (see Figure 2, Appendix B).

Sometimes it appears that farm women's concerns for the welfare of others caused some tension in the community as differing ideologies clashed. The following example about Catholic orphans points out farm women had both a practical and compassionate outlook that guided their social welfare initiatives, and enough clout to make the government listen. Florence Scissons told the story of the farm women's campaign, initiated in 1963, to change the rules about adoption in the province, to benefit the Catholic children who were left in institutions because of their religion:

> There were a lot of Catholic children in Alberta that weren't being adopted because they wouldn't adopt them into homes that weren't Catholic. And the Catholics weren't adopting them, so we in our wisdom decided that was not right, that these children should not have to live in an institution, that they should be put out into homes if possible. So we petitioned the government to change that and so we ran in to some pretty violent opposition from the Catholic Church in the form of Father O—. He thought that if your mother and father were Catholic, you should be raised a Catholic. We had no objections to that of course, except that these children were in institutions and weren't being put out into homes. So we had some discussion at the Legislature over that . . . I remember Dr. J. Donovan Ross was minister of Health at the time, and we made our objections to him . . . . the bottom line was that the ruling was changed so that people who wanted children could adopt them, whatever religious faith they were born into.

Child welfare has always been a major activity of the organization. A board member always represented the organization on the Alberta Council on Child and Family Welfare. Many resolutions requesting government action over the years were directed at the well-being of children: their health, their safety, their rights, their education and their happiness. A social welfare bulletin prepared by Mary Bentley in 1929 stressed that the "point of attack" of community improvement activities "must centre about the child."[21] The bulletin included suggestions about the formation and positive benefits of youth groups, the need to establish properly supervised recreational grounds and facilities in every community, and full support for temperance in every community. In 1979, with a Year of the Child theme to encourage them, many locals developed special projects or resolutions aimed at youth. The Notre Dame local was concerned that the revisions to the Alberta Labour Act, which were under review that year, include considerations for the protection of adolescent employees from harassment and exploitation by employers, and that meals and coffee breaks be provided in accordance with hours worked.

Day care emerged as a significant concern for women, who found that they were no longer willing to leave their children unattended while they did farm work, or for women who took off-farm jobs to keep the farm viable. Rural child care issues, farm women stressed, were different from those of families in towns and cities. Like other women's organizations, Women of Unifarm undertook a major child care study in 1986, culminating eventually in a child care pilot project shared with other farm women's organizations who were members of the Alberta Farm Women's Network in 1991. In 1990 Mary Newton expressed the feeling that more work on day care needs to be done, at the same time regretting that farming no longer provides a profitable living to young farmers, necessitating in the mother working at off-farm employment. Florence Trautman spoke from personal experience as she identified child care as one of the important things that farm women are working on:

> *Child care is one of the biggest worries. What do you do with your kids? Take them out in the barn or do you leave them behind? I lost a child leaving it behind. She was three and one half months old, she died in the house. I was outside choring. So we've got to look at that.*

Florence stressed that farmers cannot usually afford to bring a child care worker into their homes, so that other solutions, like child care cooperatives with neighbors, might have to be explored.

Environmental issues were also part of farm women's community work. At first they responded to environmental issues as they arose, not just as they affected farming as an industry, but also as they affected the quality of life for everyone. The Region 12 Women of Unifarm conference in 1977 had as its theme "This land: To use or abuse" and featured a panel of speakers focusing on farm size, land and water resources, foreign ownership of land, drainage, erosion, sanitary landfills and the "great importance of the need for preservation of agricultural land."[22] The Elk Island local reflected the concerns of most farm women when they presented a resolution to their regional conference in 1978 about chemicals used in sprays, herbicides and pesticides, requesting that "thorough research," "strong and binding legislation" and "a reliable report be given to the public . . . rather than rumours." An amendment to the motion added that these be considered with reference to "everyday use by the average home owner, homemaker or farmer" and that the "cost could be borne by provincial, federal and rural governments, as this is of concern to all people."[23] In 1988 after several pesticide resolutions were presented to government, the Farmers Pesticide Certification Course was developed.

Farm women then began to formalize their involvement in environmental issues by participating in environmental impact studies, and in the 1980s on the Public Advisory Committee on Environment. For years

Katharine Russell, a rancher in southern Alberta, was active on behalf of Women of Unifarm on the Environmental Protection Committee, one of five sub-committees of the Public Advisory Committee. In 1986 Women of Unifarm focused on the Transportation of Dangerous Goods Act as it affected farmers. The safety convener that year, Elsie Seefeldt, spoke on the subject at seven conferences and two other meetings that year, trying to make people aware of the legislation and its implications. Discussions with the ministers of Transport and Agriculture were undertaken to set up workshops around the province to inform the public of the legislation, and the Minister of Transport offered to help with technical assistance and speakers.

Farm women became interested very early in the history of the organization in consumer issues. They were interested in safeguarding public health, and concerned about fairness in advertising and marketing. These interests included the packaging and labelling of products, the identification of care or maintenance instructions, the addition of a safety seal to the packaging of food products, and the workmanship of garments, particularly work clothes. Eventually, a cooperative relationship was developed with the Consumers' Association of Canada, and a member of the farm women's group would attend provincial CAC meetings. In addition, the farm women would send their suggestions, resolutions and questions directly to the Association and occasionally include their spokesperson as presenters at conventions.

The convention of 1955 shows the broad range of interests and concerns that occupied farm women on behalf of the whole community. The Farm Women's Union that year presented their policies on health, education, divorce, property laws affecting women, mental health services and mother's allowances, and all were ratified by the convention. In addition, a wide variety of resolutions were introduced by locals on topics such as public library and school library development, the labelling of clothing and of meat, the price of drugs, the use of waxed caps on milk and cream bottles, free hospitalization for cancer patients, school playgrounds, teacher/school-board disputes, the establishment of a scholarship board and local scholarship committees, agricultural education in high school, the establishment of a provincial archives, the use of butter in cooking classes, homes for the chronically ill, amendments to the widow's pension act and the establishment of a library at Bowden Institute.

Farm women know they have made a difference in their communities and on the kinds of health, education and social services and policies developed for Albertans. Margaret Blanchard said:

I think Women of Unifarm have had a distinct effect on government. I think they are respected for their ideas and they're well received when they present the briefs. . . . And I think that's something that's been built up over the years . . . you've got to keep working at it to make sure that the ideas that

are coming forward are good ideas. . . . So I think that the influence on
government is still there. . . .

Mary Newton agreed with Blanchard's assessment: "I think there's no doubt we have made the government aware of the differences between our urban and rural situations. We've got to keep pressuring that the rural area has an equal opportunity in all these areas. . . . The government is quite receptive all the time to our briefs. There's a lot to be done. It just has to be a continuing watchdog process I guess. It always seems like a few do the work for the many, there's no doubt about that."

The women's success in addressing many issues overshadowed at times the farm men's activities, and the men felt they had to compete with the women for the attention of the media. Louise Johnston explained, "And actually some of the things that we were carrying on in the health, education and social field were the ones that were highlighted, well the media were interested, so this didn't always please." At the same time Louise pointed out that she appreciated the encouragement the men gave the women's work. She mentioned that in 1967, she was sent as the Alberta Federation of Agriculture's representative to a conference on rural sociology in Quebec where she found herself to be the only female delegate from across Canada.

Whether it pleased the men of the organization or not, farm women continued to be vocal and active in their communities and on provincial issues. As more government funding and programs became available to meet more of the identified needs, farm women retreated somewhat from their active service roles into lobbying, advisory and supportive approaches and activities. But as a forum paper prepared in 1980 (see Figure 3, Appendix B) shows, organized farm women were persistent and determined in their ongoing efforts to improve the welfare of everyone in rural communities, which included addressing both the legal status of women and women's access to services. The sense of responsibility farm women felt for this work had deep historical roots. The sense of accomplishment farm women felt from observing and remembering the changes and improvements in the social fabric of the province that they fought for, sometimes for decades, was often their only reward. All Albertans, the many who gain from the work of the few, are the beneficiaries.

Farm Women's Union of Alberta
1964 Program

Aims and objectives
1. To attain a reasonable standard of living and economic security for farm people.

2. To promote an educational program for farm women:
> To obtain improved education, health and social welfare services.
> To develop citizenship and leadership.
> To strive for international peace.

FWUA Song
(Tune : It isn't any trouble just to s-m-i-l-e)

The Farm Women's Union of Alberta's here to stay
Working hand in hand with all our men along the way
We won't give up the strife to find a better life
In FWUA

Chorus : (tune of verse)
Let's be active members of FWUA
Finding satisfaction as we work from day to day
We may be busy mothers, but we're busy helping others
In FWUA

If you're not member than we need you right away
To solve the nation's problems and to make our farming pay
So here' s an invitation to assist cooperation
Join the FWUA

WHY EVERY FARM WOMAN SHOULD JOIN THE F.W.U.A.—AND HELP TO BUILD A BETTER LIFE FOR EVERY FARMER AND HIS WIFE

1. **Economic:** The F.W.U.A. is the only province-wide organization of farm women. A prosperous agriculture means better homes, modern conveniences, and the amenities of modern living for every member of the family.

2. **Health:** The F.W.U.A. is keenly interested in the health of the people of rural Alberta. Health personnel tends to concentrate in urban centres, with a consequent unequal distribution of doctors, nurses and dentists for rural areas.

3. **Education:** The F.W.U.A. considers the education of rural children one of its major interests.

4. **Legal:** The F.W.U.A. is constantly on the alert in legal matters pertaining to women and children.

5. **Citizenship:** The F.W.U.A. affords farm women an opportunity for activity in the local, the district convention and the provincial convention. It is one of the finest forms of citizenship activity and adult education.

6. The F.W.U.A., through its local organization, gives valuable training in the conduct of public meetings. These local meetings serve as a clearing house for resolutions to the district and annual conventions.

7. **Junior Section:** The F.W.U.A. plays a large part in the success of the Juniors. These future farmers of Alberta are of primary importance. Their training in group activity will make for better farming and better farmers, better homemaking and better living.

8. Farm women, as mothers, are alive to the needs of their children, and when organized into one active association their chances of getting the best for rural boys and girls is that much greater. They realize that something must be done to make farm life more attractive so that rural boys and girls will not continue to be attracted to the cities in the great numbers that they are at the present time.

9. The F.W.U.A. is affiliated with the Associated Country Women of the World, the Canadian Association for Adult Education, the Alberta Education Council, the John Howard Society, the United Nations Association in Canada and the Indian-Eskimo Association in Canada, and lends support to many other organizations such as the Alberta Council on Child and Family Welfare, the Red Cross, the Canadian Cancer Society, the Institute for the Blind, the Canadian Arthritis and Rheumatism Society, Civil Defence, Canadian Mental Health Association (Alberta Division), Farm Forum, Farm Safety Committee, the Advisory Board of the Alberta Indian Association, the Consumers Association of Canada, and others.

Chapter Four

Sharing Fellowship

The universal benefit of the organization to its members throughout its history lies in the opportunities it presents to women to share their lives with each other. For women on isolated homesteads in the early years of the organization's existence, the monthly meeting of the UFWA was often the only contact with other women, and a cherished outing. As an early pamphlet declared:

> *One cannot over estimate the value of the UFWA to the women themselves. The stimulating social intercourse has helped to establish a more friendly and neighbourly feeling in the various communities, relieving farm life of much of its isolation and monotony. The organization thus makes possible the enjoyment of the simplicity and wholesomeness of life lived in the country. . . . Women coming to this western country a few years ago, from the East or the Old Country or the States, were inclined to spend most of their time wishing they had remained away from this lonely and sparsely populated land. Today these same women wouldn't leave Alberta to go back to their former homes for any amount; they are busy building up a strong healthy new country through the UFWA and are delighting in their cooperative efforts with other women.*[1]

This excerpt describes the blending of social needs and purposeful action that constituted the organization at all levels. They were perhaps most easily mixed at the local level, where women joined together and through meeting their own hunger for companionship with other women, they also served the community in the ways already identified in chapter Three. As interviews were conducted in 1989 to 1991 throughout the province with a wide range of women from many generations, the most common responses to the questions asked about what women remembered about their years with the organization, or what they valued most, were always the friendships they made with other women. For them, friendship was the most rewarding and enduring legacy of the organization's place in their lives. One woman described what she remembered best about her years with the farm women's organization: "the turkey suppers and the warm associations with my neighbors."

There were two ways in which women's fellowship was expressed and these were common to most local groups. One was the special sharing of experiences and fun just between the women, either at regular monthly meetings or at special events and outings. The other was the extension of their fellowship to their families and to the wider community. The 1948 program for the United Farm Women of Alberta described them as the "women of our Organization, who for many years have been the moving spirits in all kinds of cultural and social activities."[2]

Farm women describe the value they found in their women-only fellowship in many different ways. But for all women the bonds they felt with other women were important to their mental health. As Hazel Andersen explained:

> Well I think it's important that women should get together and share each other's troubles. You can't always tell your husband what you would tell your neighbor woman. And I think that's very important. And especially with kids too, talking about how to bring up your kids and that and of course you always talk about canning and cooking and share ideas. . . . Oh it's been wonderful for farm women. If somebody has a problem they bring it up at the meeting. And you find out that your neighbor has the same problem. Whether it's in bringing up kids or dealing with your husband or meals or whatever.

Even without a local group to meet with regularly, Jean Ross described her association with Women of Unifarm as a "very satisfying social outlet." Reflecting on her local group in the Cardston area, Evelyn Long said, "I never felt isolated really, but I know some of the women did. Especially the ones that didn't drive." Some locals, like the one Florence Crawford belonged to in Fenn, existed primarily for the social needs of its members. From the beginning, they did not follow the meeting programs and topics sent out from head office, and focused on their own interests, eventually disassociating themselves from the farm movement and calling themselves a "ladies' club." She identified a short period in the early years of the UFWA local when the men and women joined together for their meetings. This arrangement was short-lived because "I think the women felt that when we were joined the men held the floor; we didn't like it." The Westlock local, in their annual trips to Edmonton, focused on fun, which they found in tours, shopping and enjoying "meals in swanky places with no dishes to wash afterwards."

For many women the fellowship they found in their local group of women sustained them through hard times, and provided the ongoing support women needed to carry out the daily routine. Rita Graumans said she felt the socializing that the women of the Brooks and One Tree locals did through the years led to the women being supportive of each other, and that "being the support for one another . . . if nothing else . . . is still very important, I think, no matter how long the organization will be there." Another Brooks area woman agreed with her, as she found joining the group gave her the kind of understanding and support she could not find with her city friends, who did not understand what was happening on the farm and the pressures she experienced. She wanted something "other than machinery and cows" and she found it in sharing activities with other farm women.

The program of a local meeting had all the ingredients to make it attractive to women with a wide range of interests and abilities, at the same time uniting women by sharing knowledge, ideas and fun. A typical meeting began with roll call, designed for participation rather than just taking attendance. Women would be asked to describe a strategy for dealing with a problem of daily living, such as "how to stimulate a child's interest in good books," to share personal information that would help women to get to know each other better ("my hobby" or "what Christmas means to me") or to identify a current politician or event. The roll call would be followed by reading the minutes of the previous meeting; a discussion of current events; a reading and subsequent discussion of the information bulletins sent out by head office (for a sample bulletin see Appendix C) and of correspondence received; planning the formation of committees or strategies to deal with current projects; and the development of resolutions pertaining to issues that should be taken up at the provincial level. Meetings would often include some community singing, or perhaps some entertainment, and whenever possible, a guest speaker. They would conclude with a social time, enhanced by a simple lunch. The meetings had something for each person's taste. Many women enjoyed the social aspects the most, but many also expressed the sense of purpose with which they met together, and how this made the meetings more meaningful.

Many longtime members of women's locals spoke of the lasting friendships formed through participation in the organization, enhanced by their common interests and the "spirit of cooperation when working on projects together." These friendships endured even after women retired from farming and locals died out. Many of these women still get together regularly for companionship and to do special projects. As Jean Buit pointed out, sometimes the current members of the organization need to be reminded about the central place the social function of the organization holds for many women. She recounted:

> One of the things I really liked about being in the organization was all the people you meet. I know I was at a local that had a 40th anniversary, so they had many charter members there and several of them spoke and one elderly lady said, "I have been to many, many meetings, many, many programs, but the thing I remember most is the people I met." And I really thought about that afterwards, when we're putting on programs we're so anxious to fill up the day that there's hardly time for coffee breaks. And when you hear some of the older members speak like that . . . [it] makes you realize how important the social part of it is.

Margaret House, who joined the organization in 1931, expressed her feeling that the organization did a great deal for her personally and that the social aspect was "the enjoyable part. And that is worth everything and nobody can take it away from me." Maisy Platt remembers going to farm women's meetings as a child, with her mother, and then she became a member herself in 1940. She described the Westlock local as it was in those days:

> *It was a very big group. And it was the only organization of any kind for farm women, so it was well attended because it was more or less, . . . the only social function possible. And they just took their children and went.*

Maisy's longtime friend Jenny Stirling added:

> *And we had dinner meetings sometimes. In real cold weather we had dinner meetings. We all took the food. . . . We often had to go with horse and buggy or sleighs. There was no car then. It might be 16 miles, but it was looked forward to. . . . The men tagged along if there was something to eat. . . We still hold our meetings because we'd miss them.*

The closeness of the members of the Westlock local was evident as they shared one of the biggest activities they undertook, a three week trip to Vancouver in 1947.

> *Yes, we all got on a train to Edmonton and rented a coach and driver and went to Vancouver and we stayed three weeks. We rented a house, we took as much food as we could that wasn't perishable and lived as cheap as we could. We knew how to live cheap anyway and we had a marvelous*

UFWA meeting during the early years. No babysitters then – you just took the children along with you. (photo courtesy Jenny Stirling)

time. . . . We went to all the sites and we even went to Vancouver Island because my husband phoned and said don't come home it's snowing like fury and so we stayed another week. All of us [32 women]. Oh it was wonderful because Victoria was in blossom. . . . we all knew each other extremely well. And all throughout the years we've never had any gossip or anything.

The Westlock local thrived on a shared love of handicrafts and trips to other communities and special events, activities they still share as a group today. Although they claimed they went "seriously through" their meetings and had "brilliant discussions," it is clear the fellowship they shared in the "good social function afterwards" was what kept them together for so many years. As Jenny Stirling described it, "the social life, the friendship, has been terrific."

Maisy Platt explained, "We had dairy for 26 years, 365 days a year, 100 cows and never a holiday and I always said it [the farm women's local] was the only thing that kept me sane. Because it was such a driving thing and this monthly thing was the only relief." Rita Cannard also found the monthly meeting "a real good break to get away; I have seven children so it was a really good break. I always brought the youngest one . . . and we had real good friendships over the years. . . . I was a very shy person and I think it has brought a lot of my better qualities out."

Irene Wagstaff, a young farm woman from the Peace River area feels strongly about the social reasons for the organization's existence. "You need to get together just to have the support of someone else who's going through the same thing that you're going through, whether it's financial or emotional or whatever." Edna Butler claimed that she "would have gone insane" if she couldn't have got out to farm women's meetings, "because we never went anywhere, couldn't afford to go anywhere except to town, get groceries and come back."

Many women started going to meetings because a neighbor or a mother-in-law encouraged them. This was Sonja Hudson's experience, and for her, the social aspects of the organization constitute her fondest memories. She feels she would never have met some of the women she calls friends if not for the farm women's local. Although she has enjoyed conventions and the learning they have brought, making friends who have the farm background in common has been the most rewarding outcome of her membership. Naomi Findlay also was invited by her neighbor, Mrs. McIllop, who was a director of the Farm Women's Union at the time, to join the organization. She made it clear Mrs. McIllop would not take no as an answer. She also expressed the feeling that the "fellowship of the ladies" was more significant than anything. Louise Christiansen explained her motivation to join:

All my friends were going and my mother-in-law was going to meetings and so I went as a social, it was a social event. You got out once a month.

> *At that time there wasn't that much going on in our small community, so it was a time to socialize.*

Getting the household organized and the farm chores done in order to be free to attend a meeting was quite a feat. Olwen Murray of the Heath local described her efforts to attend her first farm women's conference in 1943:

> *I was one of the few young women and mothers in the area and my neighbors and friends thought I should go to the conference to learn about the farm organization. So I decided to go to leave at 8:30 a.m. I was always up at 5:00 to help with the chores, milk cows, feed chickens, look after the hens and help pump water for the horses. Then breakfast, washing the separator, and two small children, one two and half and the other six months. I thought I was doing well and rushed six miles to the home of the driver for the day at 8:35 a.m. When I got there the other four ladies were waiting and the driver says, "You're late, don't you know you're supposed to be on time when you ride with others." How I wished I was home. It was a great conference at the home of Mr. and Mrs. Spencer. Mrs. Ivy Taylor was the guest speaker. Since 1943 I have attended more conferences than I have missed.*

Music and community singing were often key parts of programs at both the local and provincial levels. The fellowship that grew from people singing songs together, preparing musical entertainment for each other's enjoyment and relaxation was fondly remembered by many women. Janet Hogg, of East Longview, was appreciated by many women of the High River area for her years of faithful piano playing at all the farm women's meetings and conferences. Women also enjoyed discussing and looking at photographs or slides of other countries, sharing their own travel experiences or those of special guests.

Handicrafts were a major focus of the time many women spent together at the local level. The handicrafts competitions at both the regional and provincial conferences were for many farm women a highlight of the gathering, and many worked hard to produce beautiful sewing, knitting, embroidery and crafts to submit for adjudication. The job of organizing the handicrafts competition and display was a huge one, and involved many long hours for the Handicrafts Convener and her assistants. Molly Coupland was Handicrafts Convener for many years, and as Mary Belanger who assisted her explained, Molly and her assistants were unable to participate fully in convention business because they were so busy in the crafts room. The list of categories for entries to the craft competition found in Appendix B, Figure 4, demonstrate what a large responsibility the handicrafts convenership had become over the years.

Although some locals, like Westlock, maintained their love of crafts and continued to make it a part of their local activities, this focus died out over the years, and the official competition at the provincial convention ceased in 1985. The lack of interest and time for craft making is cited as the major reason this aspect of women's fellowship disappeared, although some farm women set up individual tables of their own work at annual conventions in the late 1980s and into the 1990s. The Westlock local was pleased to find that there were young farm women in their community eager to carry on the local's craft traditions; however, most locals find that crafts became over time a personal rather than a group pursuit. Women still appreciate the fine work done by other women, and crafts produced for specific purposes, such as a quilt or an afghan, to raise money for a good cause, or to celebrate an anniversary or other special event, are still popular.

Friendships at the local level brought a spirit of sharing and caring for each other that extended to members' families. The Freedom local's members for many years always helped each other by catering at cost for special family events, like weddings. Teresa Maykut expressed her feelings about it: "I thought it was such a nice idea, even though you worked as hard as you did at somebody else's. But it was such a small little group that you wanted to give to each other you know. Just the closeness there."

Many social events were planned to include the men's local and the husbands of women in the women's local. The High Prairie local focused much of their energy on social events, like curling bonspiels, teas, dances and bake sales. As Inga Marguardt described the bonspiels, "it was a real big event. We had big pots of home-made soup and chili and the men all enjoyed themselves and the women did the work as usual." Besides the fun of teas, bake sales and dances, they were "the only way we had to make money at the time," Inga explained. And this money was usually used to meet a community need. In High Prairie, the women helped to build the museum, the first library and "always contributed to the library." Inga says her fondest memories are of the Farm Union picnics and the curling bonspiels, and that for her, mainly in the 1950s and the 1960s, the farm organization was important for its social activities, but that in the 1980s and 1990s "farm social life is falling apart."

Greta Hallet remembered the plays the Fleet local put on for the community and other Alberta towns. Picnics and box socials were also favorite activities of the Fleet local. The One Tree local described the socials and dances that were common in the early years, and that in fact they danced so much they had to put a new floor in the Southern school. The restroom they built in Brooks became a community centre, for activities like card parties, teas, showers and Christmas concerts. The One Tree local also gave a cookbook to each new bride.

The Dakota local held a community picnic for many years, which they eventually expanded to include the whole county. They arranged for the district agriculturalist to give a tour of the county, and the farm women provided the picnic dinner at noon. They were active in the community centre, held annual wiener roasts and always organized a Christmas party. Every year the Grandmeadow local held a turkey banquet for all the farm families, exchanged cards and visited. The Eastburg local became known for its musical skits, often put on at regional conferences for other farm women. They held a bazaar every fall, including a bake sale, held dances and farmers' picnics regularly, and they still get together periodically for potluck suppers.

Miriam Galloway, of the Fort Saskatchewan local, said the best memories were of the "fun things, the social times when neighbors got together; we had our summer picnic and special meetings with special speakers. . . . it gives a support group to us, not just when we're needing support, but all the time. It gives opportunities to share similar problems." Miriam added that the organization was "dear to my heart. Just having opportunities to be with my neighbors, it's the only time you see the neighbors sometimes. And sharing friendships." When asked what the organization meant to her, Dorothy Ottewell, also of Fort Saskatchewan, replied, "It all seems to be pleasure in my mind because a lot of it was talking to other people when you're at a convention or a conference. I met a lot of people I would not have met otherwise." Dorothy added, "It was one way for new people coming into the area to get acquainted with everybody too. . . . We would go and invite them to come to our meetings. Some of them were glad to come, others just looked at us, but on the whole they came." Dorothy felt the "entertainments" that the local held were a significant contribution to the community "because everyone was invited. And we held them in the schoolhouse as a rule. The card parties were in individual people's houses. . . . But we used to have so many dances in the old school house; it was lots of fun." Marguerite Mason said she thought their "Christmas parties are just fantastic. We really enjoy them. We just have such a good time and like my husband said, he would never have got to know all these people if it wasn't for Unifarm. That is a very fond memory. Another good memory I have is when we catered to auction sales. We had such a good time and you saw everybody. . . . I enjoy all the things we do together."

The Sunniebend local's social activities always operated around a hall they built. They would hold dances every two weeks and they'd have chicken suppers and whist drives and shows. Doreen Sexty says the Sunniebend local has "made a great community bond amongst the women. And it's an outing they all look forward to." The Sunniebend women work together with the men on pancake suppers, picnics and the annual Christmas party. Yet Doreen felt strongly that despite the continual emphasis on social activities, that "you have to have a purpose to meet. You can talk about all the social stuff you want, but you've got to have a purpose and if you don't have any reason to meet it will fall flat. So we use the women of Unifarm as our objective and then our social stuff is thrown in."

Emma Innocent felt that Women of Unifarm and its predecessors "holds a community together" because "you've all got a common interest all the time." There is a real connection between the relationships that women build with each other and the success of their political and social reform projects. Janet Walter suggested that what the farm women have accomplished for the community at large is "complex and it was very useful to the social fabric, because it's the showers that welcome the people getting married, and the baby showers and the wedding receptions and providing a lunch after a funeral. The whole range of human experience, they are the background that provided the milieu for it [social reform] to happen."

The catering services that farm women provided for special events in their communities were opportunities both to raise money for community needs and to have a good time together. Many women spoke with good feelings about the suppers, lunches and canteens they prepared and served, providing an important community function while enjoying a special kind of fellowship. In many ways, these food service activities kept farm women connected to the larger community and in tune with the spectrum of community interests and needs. This knowledge of and connection to the interests and concerns of other groups helped women, through the traditional female role of serving food to others, enhance their political effectiveness in the community. Many ideas for political and social reform arose out of interactions at auctions, fairs and suppers.

Although most of the community social functions were directed at fund-raising, the organization rarely benefited from the profits it made. Money was channelled back into the community, into charitable projects and organizations, into the financing of needed services or facilities, or

Fund raiser at the Berwyn Farmer's Market. From left to right: Zella Pimm, Betty Ann Schur, Netta Connolly, Mary McKenzie, Judy Pimm. (photo courtesy Judy Pimm)

into helping particular individuals in need. Farm women used the resources they were best able to put their hands on, the production of food,
the making of crafts, the productions of musicals and dramas, to build
better communities. Through doing what they knew best, and using what
many women described as the only means they had of making money
for many decades, women accomplished a two-fold mission. In a seam-
less interweaving of content, function and intent, farm women contributed materially and spiritually to the lives of their communities and also
contributed spiritually and emotionally to their own well-being and that
of their own families.

The wide range of activities carried out by locals is impressive. Farm
women displayed, throughout the generations, a talent for organizing
social activities that cost little but rewarded many with meaningful
fellowship and lasting good memories. Food was often at the centre of
the activity, as women knew that food was an attraction for young and
old alike, and provided a special kind of sharing. Pot luck suppers and
picnics were popular with many locals, for everyone contributed food,
and the focus was on getting together. Teas, bazaars, amateur nights,
musicals, dances, including square dances, and plays were also popular,
though planned with less frequency because of the work involved.
"Many a play we put on and even travelled with our cast to outside
points," wrote a member of the Ridgewood UFWA.[3] Almost every local
held a Christmas party, most included a turkey supper. Members often
exchanged cards or gifts. Bingos, whist and other card parties, including
telephone card parties, were enjoyed in many locals, and were often a
source of fund-raising. As one member of the Ridgewood local described
them, "these card parties [were] one of the best community projects we
ever had, as young and old gathered every two weeks in the hall and
played whist. At the beginning there were many hired men in the districts
and the card parties gave them some place to go and meet people. These
card parties were carried on for years."[4] Special events were planned to
mark important happenings in the community. The Brooks local decided
to have a party to celebrate the local's tenth anniversary in 1938. The
Ridgewood local organized a banquet to celebrate the return of the
town's men from the war in 1945. In 1946, the Horn Hill local invited
Ridgewood to join them in planning a "banquet for our husbands."[5]
Showers and birthday parties were held for members and their families
in many locals. Twenty-fifth wedding anniversaries of members were
usually commemorated in some social function. Wiener roasts, skating
parties, strawberry festivals and ball games were all mentioned as social
activities organized by the Notre Dame UFWA in the 1930s.

Women's groups often hosted each other throughout the years. A
neighboring local might extend an invitation to another. Sometimes
Women's Institutes members in the same or neighboring communities
would be asked to come to a particular social event, or be included as a
partner in organizing one. Often a successful event would become a

tradition, organized annually. This was the case for the Ridgewood local, which included women from Pine Hill and Ridgewood in its member-ship.

> *Our first money-making venture was an Irish stew supper held in the Hall on the 17th of March. The roads were bad; people came in wagon-boxes and democrats and some drove through mud holes so deep the water came into the wagon boxes. However, everyone came, enjoyed the supper and the concert put on by local talent. From then on we had Irish stew suppers for years, also chicken suppers in the fall.*[6]

Chatauquas entered the province of Alberta largely through the en-ergy and enthusiasm of United Farm Women members. The Pembina constituency conference at Freedom in 1937 proposed that a chatauqua be held instead of a conference, and the first chatauqua in Alberta was organized by the United Farm women and men that year. It was a huge success, and was repeated for several years. Farm forums on the radio, broadcast on Monday evenings throughout the 1930s, was another op-portunity to connect with neighbors, as everyone sat around the radio to listen. Not everyone had a radio at first, so sharing and discussion was a part of the evening for many farm folk. The women regarded the broad-casts as important enough to set aside their chores for the evening, and to assign a member of their local to listen and take notes for discussion at the next meeting of the local, just in case all the members did not get a chance to be near a radio.

One aspect of fellowship that was carried out, and in some communi-ties continues to be attended to by farm women, is the acknowledgement and comforting of people who were ill or bereaved through gifts of flowers, fruit, baking or cards. While these gestures seem small, they were not insignificant, as community members felt a sense of belonging as a result of these efforts. The Clover Bar local felt this was an important part of their work in the community, and their list of recipients included many who were "shut-ins" most of the time. They also prepared Christ-mas boxes for needy families in their community, and took candies to the elderly throughout the 1930s and 1940s.

The relationship between farm women and the care of the elderly has a long history. Being the first organization to lobby for lodges for elderly citizens, farm women have maintained their interest in all aspects of elder care and lifestyles. Most locals participate in regular visits to lodges and nursing homes, offering entertainment and companionship, hosting birthday parties and donating fruit and baking.

Many locals also made an effort to recognize Armistice Day, later Remembrance Day, in meaningful ways. For some communities, the program planned by the farm women's organization was the only public recognition of the day's significance. And during war time, farm women extended their sense of community to include the men and women overseas. In addition to their "war work" of knitting, sewing for and sending money to the Red Cross, they also sent parcels of books and treats to the men and women from their communities, and held numerous fund-raising events for the Prisoners of War Fund. The Ridgewood local, for example, at Christmas in 1942, sent a box of chocolates and handkerchiefs to the "boys in uniform" from the community who were still in Canada, and Christmas boxes to the four men from the community who were overseas.[7]

There is no doubt that for many years the farm women were at the centre of the activities designed to build neighborliness and fellowship in farm communities. The Ridgewood local is a good example. They summed up in their 1929 year-end report that they had organized "12 whist drives, two suppers and concerts, two plays, one sewing demonstration, and that flowers and fruit were sent to 15 people."[8] While the farm women were the primary social organizers in many communities in the first decades of the century, this role changed gradually as other community groups were formed. In the 1950s, for example, the One Tree local was still active in setting up the schoolhouse as a community centre to ensure that a centre for children to play and for adults to square dance was available through the winter. The restroom in Brooks, a project of both the Brooks and One Tree locals, remained a centre of activity well into the 1960s, when the Town Council decided to destroy it. And the One Tree local continued to sponsor anniversary showers and going away parties well into the 1960s. Farmers' Day picnics and Christmas parties were organized by the Readymade local for many years, and were still popular with members and their families in the 1980s. Mother and child picnics, popular in the 1960s, were replaced in the 1980s by a mother's day luncheon by the Readymade members.

Although the types of activities changed over the years to reflect changes in people's interests and resources, farm families continued to enjoy opportunities to socialize together because of the initiative and planning work of farm women. There is no doubt that as the farming population decreased in numbers, and the availability of a wide range of recreation and social activities became available, that the central role farm women played in organizing community fun diminished. But among the members of the organization and their families, the picnics, suppers, dances and parties still remain favorite and memorable events.

Beyond the local level, there were several ways in which women experienced sisterhood and connection with other farm women. One was at Farm Women's Week, another was by working as a director on the provincial board, and the last was at national and international forums, such as the Canadian Federation of Agriculture, farm women's conferences and the Associated Country Women of the World. Although the numbers of women involved in activities at these levels were small, nevertheless the contacts women made and the friendships that grew from them were regarded by many women as the most meaningful experiences of their lives.

July 16-19, 1962. Farm Women's Week, Olds Agricultural College.

Farm Women's Week was initiated in 1930, the inspiration of a farm woman named Isabel Townsend. Held at the Agricultural College in Olds, Alberta, it was designed as a week of education and relaxation for farm women away from the pressures of farm life. Over 30 women attended the first Farm Women's Week, and its popularity continued for many years. Later renamed Alberta Women's Week, the program was cancelled in 1986. A second opportunity for women to attend a Farm Women's Week was introduced at Vermilion School of Agriculture (now Lakeland College) for a period of time. For a number of years, the provincial board of directors ensured that two board members attended the two Farm Women's Weeks, to help the organization keep in touch with its membership and to bring the information shared in educational sessions back to the board. As Mrs. Banner recommended it to the farm women at the 1941 Region 11 conference, Farm Women's Week was "a good way to have a holiday."[9] Spending a whole week in activities together gave farm women a chance to make new friendships with women from other parts of the province, friendships that would continue through contacts at conventions and other farm activities (see Figure 1, Appendix B).

Many women found deep satisfaction in the social contacts they made at the provincial board level of the organization, and it offered a particular kind of exposure to and sharing with people from all over Alberta.

When asked what were the highlights for her in all her years on the board of Women of Unifarm, Mary Newton replied, "Oh, the people . . . you do meet people from every end of the province . . . often now if we drive anywhere I think, I could stop here and see so and so, and you could just about tour the province and visit. And that's a real warm feeling, you know, people are probably the number one thing of it all."

Many women who were able to participate at the provincial board level cite the opportunity to meet women from all the different regions of the province as one of the best experiences of their lives. They enjoyed the ways their lives were enriched by the exposure to different points of view, and to women's diverse problems. In a very tangible way, the social exchanges with other women were the best education the organization offered. Louise Christiansen, who eventually became president of Women of Unifarm, stated that at the local level there was "such a feeling of belonging," but also as she became a director "it was the socializing with women from all over the province that I think are really fond memories. We had get-togethers in the evenings and really had a lot of fun together." She added, "I guess it's given me a feeling of well-being and it's broadened my views on how other people live. It's a satisfying feeling I guess." Verna Kett agreed that it was "the people themselves" that stand out for her as the best part of her involvement and the fact that "you know that you've got somebody who really cares." She also feels involvement in the organization "builds your self-confidence and your ability to communicate with other people." Sometimes too, "just to be away from the farm" gives one "a different perspective again." Like Mary Newton, Verna feels she could travel the length and breadth of the province and know somebody in most of the towns, and that if she was to need friendship, that she would get help. . . . "the directors, they really do know you and they care. It's a real kinship and I wouldn't give it up for anything." Joyce Templeton, a regional director for Women of Unifarm, said her fondest memories are the "associations with the farm women at my local, but also the association with women at the provincial level. I think it has expanded my thoughts." Janet Walter, also a provincial board member, believes that what the organization has accomplished for farm women is "thousands of friendships, thousands of the kind of support they needed, and that support is as varied as the needs of the people involved, but I'm sure it includes what many people get from paid therapists."

The Associated Country Women of the World gave some Alberta farm women a unique fellowship with women from all over the world. While some women gained an increased understanding of the situations in which women from other cultures found themselves, some women also developed lasting friendships with women from other countries. These were maintained through the exchange of letters and sometimes visits long after their initial meeting at the world convention.

The contact with other women that women prized so highly proved to be an education, a source of new understanding and personal satisfaction. At the local level as well as the international, women expressed the belief that being part of the organization has broadened their understanding of other people and their troubles, and that this compassion for each other "held a lot of farmers together when things were tough." Several women of the Edgerton local felt that the best part of their many years in the organization was "going to conferences and conventions because we always learned from others. . . . opinions and ideas." Barbara Klymchuk expressed her appreciation for the mix of people that Unifarm brought together. " I think that's one of the wonderful things about Unifarm is that we're all Canadians and we draw members from every race; it's just delightful to go to a meeting and see people from all over the province, all different walks of life. Some are well off and some are not so well off."

Dorothy Ottewell expressed the belief that participation in the women's local in the early years helped women to get out of the house, to go and see people and visit at the time of the meetings, because "at that time when we were farm women, we were isolated. And if you didn't drive a car, you were even more so." As Corinne Thompson, who joined the organization in 1985, pointed out, although the "social aspect" was the reason why many of the women joined the organization in previous decades, that reason no longer attracts new members to the locals, "because people don't need to have that social void filled" anymore; there are so many other activities going on in rural communities. Because of her participation at the provincial level as a director for her region, Corinne has made many friends across the province and developed an understanding of their different perspectives. But Corinne does not have a local operating in her community. Rita Cannard agreed with Corinne Thompson, that the needs of farm women have changed. She described it in this way:

Well, years ago it served a real purpose for the ladies; we'd meet perhaps once a month and that was the only contact with the other ladies to talk about their families or health, so that was really a need, but now I don't think that need is there any more. Young people are too involved with their children and with other things.

Still, Rita Cannard felt that the social needs of existing long-time members are central to the organization at the local level. She pointed out that though locals do not traditionally meet over the summer months, members of the Westlock local feel it is too long between June and September to wait to see each other, so they plan a luncheon over the summer, or a trip to a museum with a picnic lunch.

After conversations with many generations of farm women involved in the organization, one quickly learns that appreciation for the lasting friendships and the caring fellowship of other farm women is central to their experiences. Despite the worries of contemporary farm women that the social life of the farm community is disintegrating, farm women of every generation have managed through the organization to create and maintain meaningful human relationships that they cherished and from which they grew as persons. This is an accomplishment for anyone living in a society that is increasingly fragmented and individualistic. Farm women have built supportive communities for themselves among their own membership, and extended that sense of community to their families and their neighbors. The strength their networking and their friendships have given communities is irreplaceable. And the warmth and meaning their connections to each other have brought to their own lives are both priceless and enviable.

Women of Unifarm Objectives
1970

To work as farm women, cooperating with the men to attain the objectives of Unifarm;

To study existing legislation, and proposed legislation particularly in its relation to the well-being of women and the family;

To obtain improved educational, health and social development services for the farm family and the total rural community;

To formulate and promote provincial and national policies for the protection and well-being of family life;

To promote a continuing education program for farm women;

To develop citizenship and leadership in rural women;

To strive for international peace through affiliation with Associated Country Women of the World.

Chapter Five

Citizens of the World

One outstanding feature of this organization throughout its history is the way in which it empowered farm women in the exercise of their citizenship. Women educated themselves and each other on a wide variety of important political and social issues. These educational opportunities were combined with the efficient use of the organization's structure to influence decision makers. This process was often effective and profitable in outcomes. At the heart of this process was the farm woman herself. Whether motivated by concern, anger, curiosity or frustration, she engaged herself in social problems and their potential solutions, took the time to do the research, to talk to those affected and to understand the issues. And then, perhaps timidly at first for some women, for others with a feeling of power that knowledge and indignation supply, the problem was named, the ideas were put forward, the strategies were planned and the actions were undertaken. The organization taught each farm woman how to be a social activist, but more importantly it taught her that she had the right and the ability to be one. This is a most powerful legacy of farm women's organizations. Farm women who often felt isolated and overwhelmed in their own lives found in the organization a place where they met their social needs and where they also learned that an individual can make a difference. They learned that both individually and collectively they could not only hope for change, they could make it happen. A speech by Lucille MacRae, who became secretary in central office to the United Farm Women in 1925, suggested "Farm women are all, in a sense, business partners, with common problems and interests welding them together, naturally, more closely than any other class of women on earth. For that reason, they should lead the world in organization work."[1] Whether farm women felt this expectation for their collective success is not clear, but they collectively maintained over the decades an admirable record of active citizenship in the province and in the world.

Fundamental to farm women's view of citizenship is a sense of entitlement as well as a sense of responsibility. In an industry and a lifestyle dominated by men, but dependent on the work of women, farm women learned quickly that they would have to assert their entitlement to a voice, a vote and an occasional vacation. And farm women did act responsibly in using their voice and their vote. They also learned how to turn political and community work, demanding in its own way, into a kind of vacation away from the rigors of farm life. The kind of citizenship they demonstrated had many expressions, including personal development, outstanding leadership, creative money management, their extension of human service to the world community, their leadership in working with young people, their efforts toward world peace and international cooperation and understanding, and their vision of cooperative networking among farm women at the provincial and national levels.

Personal Development

Many women reflected on the personal benefits of their participation, and talked about the outstanding leaders whom they grew to admire and respect. Farm women identify the personal growth that they experienced as they became involved in activities beyond their own farm gate. Florence Trautman reflected on her own changes after joining the Women of Unifarm Board:

> In this last year and a half I've had comments [from people] and they've said "God you've changed in the last year. I have more self-confidence, I would have never ever spoken in front of people before. . . . I never knew I could do what I'm doing now. And I give credit actually to Women of Unifarm in an awful lot of ways . . .

Maisie Jacobson also found her years serving the farm organization brought her personal rewards:

> It has contributed greatly to my personal growth. As I became interested in betterment for rural people, people in general, I found that my horizons widened. My caring increased and my awareness of what was taking place around me became greater and the thirst for more knowledge and the striving to find new ways of helping people, making it a better world. I found it to be a real challenge. . .

Louise Christiansen shared her belief that one of the accomplishments of the organization is that it "brings out leaders" by giving women an opportunity to develop leadership skills. She felt that in her local, "more of the women feel at ease acting as chairman now than they ever did." Jenny Stirling pointed out that all the members found themselves holding office of some sort, and that "in our day, we didn't have the training and to stand up before people was a real effort." Rita Cannard said that her participation in the organization had taught her leadership skills. She also felt the exchange with Quebec women in 1975 and 1976 were important educational opportunities for the women involved. Corinne Thompson indicated that she had learned a great deal about the education system and the health system of Alberta through Women of Unifarm. The opportunity to attend the Canadian Federation of Agriculture annual meeting really increased her understanding of agricultural politics and the role of the provincial organization in it. Margaret Blanchard, a president for two years, shared her perspective on what she gained:

You would never get paid financially for what you do, the hours you spend, but you could never buy that education for any amount of money. You could never have the experience of meeting all the people and seeing all the country and you know you couldn't buy that kind of education, if you went to university for 20 years you would never get it. I think that to me has been the real advantage to having been able to serve.

Some farm women also saw their involvement in farm organizations as contributing to their children's development and to a better marriage. As Joyce Templeton described it:

It implants in your children the value of a community, and the value of working together . . . I think it instills in your children too that you have a debt to pay to the community. . . . doing things that interest you certainly broadens your children's minds too . . . I think it instills in them wanting to do things too, to go places and learn more about all the concerns of the world . . . I always think it makes a better marriage when a woman gets out and broadens her horizons a little bit. And it makes you more broad minded. . . . farming is a business and if a woman doesn't keep up with the issues that come out of the farm organizations, well, I think you've lost a lot in life. And that's one way of keeping up with the issues, by joining a farm organization like this.

Leadership

The leadership of the organization over the years inspired women who joined it to express their citizenship. Individual women provided by example the kind of commitment and determination that inspired other women to participate whole-heartedly. They built into the organization the belief in the necessity of cooperative and intelligent social action. An accomplishment of equal value was the ways in which they lived this belief, and made it the foundation of their approach to working for social betterment on a daily basis. To modern women, with access to resources unimaginable to farm women of yesterday, these achievements may seem inconsequential. But for women who found that getting to a monthly meeting was at first problematic because of the pressures of family and farm work, the commitment to any form of activism, even the task of researching an issue paper and presenting it to neighbor women, was formidable. And many women went far beyond this beginning point.

Some leaders were not well-educated in the formal sense, but they undertook a self-directed education of staggering proportions. Partly out of a belief that they must do their very best, and partly from a feeling of insecurity about how much they knew, women became very well informed on a wide variety of topics and issues. When they were asked to be responsible for a specific area, they undertook this responsibility conscientiously, producing informative and well-written reports, briefs and bulletins. If it seemed to some that the cabinet ministers and premiers took farm women's views seriously, there was good reason. Those women were as well-versed on the issues under consideration as the ministers themselves, and their life experiences sometimes made them more knowledgeable than the honorable members of the government.

Almost every farm woman interviewed to produce this story singled out a particular farm woman who had influenced her in some way to be involved and for whom she felt a great deal of respect and gratitude. Women recognize the capabilities and commitment that made some women outstanding leaders and mentors for other women. Some of the women who were mentioned served as presidents of the organization, others were district or regional directors, some were neighbors who kept their locals strong and active for many years.

Of the many wonderful qualities these leaders and mentors possessed individually, the one common to all was commitment to a vision. Organizational work meant more to them than just keeping things running smoothly and meeting the needs of the community. They regarded the organization as a powerful vehicle for social change. They understood that to remain powerful, its membership required continuing education, inspirational thought-provoking communications and fearless, dynamic leadership. These women were not content with the status quo. They sometimes showed a restless impatience with those who were unable to grasp the possibilities they themselves envisioned. They were also not content to do all the work themselves. These leaders held other women to the same standards of service and commitment that they followed themselves. This required the sharing of leadership. One member talked fondly of how the director in her area "cracked the whip" and everyone was glad she did, for together they accomplished a great deal. Another woman described another leader, Hazel Braithwaite, as a very able and energetic individual, who had ideas that came to her "just like kleenex popping up out of a box: she never ran out. So it was certainly a pleasure to work with her, just one of the few treasured memories that one has in a lifetime."

Leaders who shared their leadership empowered other farm women. It strengthened the organization, as new leaders were prepared to take the reins from the old. Betty Pedersen talked about grooming the vice-president who worked with her while she was president, and stressed that the president's job was one that took time to learn, sometimes several years. "You don't just walk into it.... There has to be a period of growth." Mabel Barker emphasized the requirements for directors in her day, and it was clear she saw these as important to maintaining the whole structure of leadership and representation. She spoke of the many hours committed to travelling and keeping in touch with people personally, helping locals deal with their particular challenges.

For many years, directors of the organization and any interested members were offered leadership training courses at the Banff School of Fine Arts in winter. Sponsored by the Rural Education and Development Association, these one- and two-week opportunities stressed training in policy development and public speaking, and were held jointly with the men.

Farm women were cognizant of the need to establish credibility with the men, both within their own organization and in government, in order to be taken seriously. It was a case of doing their homework and learning "the language" that men understood, the language of farm economics. As Betty Pedersen pointed out:

> *You had to prove yourself before they would accept you. Once you had proved that you had studied whatever problem they were working on and had a grasp of it and knew what they were aiming for, then they would accept you. . . . You had to talk their language and they wouldn't accept you unless you did. . . . at times I found it difficult.*

One of the examples of the kind of leadership the organization engendered was the willingness to tackle every issue of importance that came along, quickly and with intelligence and a commitment to see it properly addressed. One of those opportunities mentioned previously was the Murdoch case, a specific example of the unfair division of property between a divorcing farm couple. The organization had been pushing hard for community property legislation for decades, and were frustrated with the lack of action and support for new legislation. They responded to the plight of Irene Murdoch by hiring Bill Gill, a Calgary lawyer, to appeal her case in the Supreme Court. Following the shock of the unsuccessful appeal, the organization persistently worked on the issue until the change in legislation they had sought for decades was introduced in 1978. As Helen Murray remembered it:

> *I was director in this area when Mrs. Murdoch had problems. And I let Betty Pedersen [President] know about it. And of course she knew Bill Gill in Calgary, and he was the lawyer that really helped things and he helped all the women of Canada. It rocked Canada, but it had to go to Supreme Court, Alberta would do nothing for farm women. And this was a little disgusting, because as farm women have 11 jobs, and at that time you could not pay your wife wages, and it was real impossible to ever get joint property.*

Betty Pedersen gave another example of the women's leadership in responding to issues as they arose. She mentioned that she, like other presidents before her, would read every inch of the newspaper and if there was any issue about which she felt the farm women should provide some opinion or information, she sat down and wrote a letter or a brief immediately, consulting with other board members in the process. This ability and commitment to respond and the development of broad-based leadership began to slip away in the later years. As the farming population aged, and its size decreased, and as new organizations and activities competed for the precious free time of farm women, the concern for the viability of the organization did not translate into nurturing new leadership. Leadership courses were offered throughout the years, new members were encouraged to join, and many women demonstrated the personal growth, in skills and self-esteem that organizational work brought to their lives. However, while individuals gained, the organization lost when these individuals moved on and there were fewer and fewer women to replace them.

Money and Membership

In a self-governing organization, the size of its membership and the size of its bank account are inextricably linked. A group blessed with many contributing members is strong both financially and in human resources. With a decreasing farm population, and some difficulty in attracting young farm women to the organization, the issue of finances has become an important one in recent years. From the beginning, however, farm women have shown a resourcefulness about accomplishing a great deal with whatever material and human resources were available to them. It is something they learn on the farm, and that they transfer easily to their organizational work.

One of the most interesting aspects of the farm women's movement in Alberta was their approach to money. They demonstrated a conscientiousness about money rarely found in many organizations, perhaps because most of the financial resources of the organization came from their own pockets and their own labor. Money was a means to an end. Alberta farm women were not interested in having huge bank accounts, at either the local or provincial levels. Fund raising at the local level was always undertaken with what the funds could do for others in mind. The talent they displayed for making money from the simplest activities has already been discussed in chapter Three. Money was an expression of citizenship, because it represented an opportunity to make changes and improvements, to make a difference. It was never a source of power for farm women, it was instead a source of empowerment that was responsibly and compassionately shared with the communities in which they lived, and applied to the causes that they supported.

At the same time, finding sufficient operating funds for the expenses of the executive and the board, and to conduct the business of the organization was always difficult. In the early years some of the board members paid their own travelling expenses. Later, as Mabel Barker described, directors were paid ten cents a mile for gasoline to cover their districts, making their annual trip to each local, and some additional trips to conferences and invited appearances. Originally, no member of the organization was paid for the hours of work she did on behalf of organized farm women. In later years, the President was awarded an honorarium to compensate for the many hours her job demanded.

One of the best money makers, and one that became a tradition over the years, was the organization's cookbook. First produced in 1928 from recipes donated by members from all over the province, the organization published the book themselves, with assistance from Co-op Press. In total they produced eight editions over five decades. The organization sought a publisher for the eighth edition in 1989 and produced a new, updated version of the cookbook, complete with microwave recipes and information about environmentally friendly cleaning products. Although some long-time members were concerned about giving up the publication rights to the cookbook, it once again proved to be a popular publication, and the royalties have provided the organization with much needed funds. Another method of raising funds for the organization was through the sale of souvenir spoons, a popular collector's item for a number of years.

The other major source of funds was from membership fees, channelled from the parent organization in the form of operating grants for many years. Some women expressed their displeasure at this arrangement, as it put the women's organization in an inferior position, always having to ask for money from the men. Constitutionally, the women were never established as an independent entity until 1991, and so always operated trust accounts of various types for special projects and scholarships under the authority of the larger organization. This led to disputes over the years, as some women resented the control the men had over the finances and disagreed with some of the decisions that were made. Eventually, this led to a desire to develop a separate constitution for Women of Unifarm in 1987, and to set up a separate bank account for Women of Unifarm. Margaret Blanchard, who was president in 1987, led the organization through the decision to make this change:

> *The more independent you are, I think the better off you are, the more your voice will be respected. I'm concerned that women won't hang onto that independence, that they'll become dependent on funding. I guess I was really instrumental in arranging that we take our constitution and bylaws out of the Unifarm constitution and financial matters out of Unifarm and into our own bank, because I feel that if women don't have control of their own funds then they won't feel that they are really in control of their organization.*

Despite the moves toward independence, however, Women of Unifarm still did not achieve legal status as a society or financial independence even after the new constitution was ratified in 1991.

In 1989 Women of Unifarm decided to conduct an internal review of the organization's structure, membership and decision-making processes. Many factors contributed to the need for a review. The number of functioning locals was declining and two proposals had been put forward in the 1980s, one by the Pelican local and another by the Blindman local. The Pelican local asked to change the membership requirements for Women of Unifarm, to admit interested farm women as individuals to Women of Unifarm. Women would be free of the necessity to be part of a "unit" or "family membership" in the larger organization of Unifarm in order to participate in Women of Unifarm. The Blindman local asked that Unifarm memberships be issued to individuals, at half the cost of unit memberships. When in 1981 Region 10 put forward the first request to have women admitted to Women of Unifarm who were not members of Unifarm, the Board refused to support the resolution. In response, farm women of Region 10 reorganized their regional structure and membership requirements to admit all interested women, and set up their own committees. They still sent a director to Women of Unifarm Board and participated at convention in the usual way. However, it was clearly important to the women of Region 10 that a more inclusive membership policy be implemented. The internal review, conducted under the leadership of John Melicher of Rural Education and Development Association, and completed in 1990, led to two changes. One was a resolution put forward by Blindman local that reflected a strong feeling among many farm women at the time, and that was a plan to dissolve the women's organization and encourage women to continue to participate as members of Unifarm. After heated debate and much soul searching, the 1990 convention delegates voted on the resolution to dissolve the women's organization and came up with a 50-50 split on the vote. As the dissolution of the organization required a two-thirds majority vote, executive members were then charged to carry on the business of the organization as usual. As a result of this process and vote, some members left the organization. Those women who remained faced the challenge of rebuilding, and they approached it with a positive attitude. Elizabeth Olsen, president during this critical period, commented on this process:

> Last year was a very stressful, inward looking year and I thought the end debate was exceptionally good. The quality of the debate was excellent and people got up and said what they felt ... I really felt that after we had finished there was no animosity there ... it made us who believe really strongly in the organization feel that we had something to build on and expand on.

The second change that resulted was a new constitution for Women of Unifarm, which was ratified in 1991. However, the requirements for membership were not altered. Women still had to be members of Unifarm before they could participate as Women of Unifarm members. Women of Unifarm established its own bank accounts and trust funds under this new constitution, and arranged for an annual grant from the Unifarm membership fees of ten percent of membership fees.

Women of Unifarm were not solely dependent on money from memberships and cookbooks for their activities. For special projects, such as issues studies, the farm safety hike, the child care study and others, the organization received government grants, usually from Alberta Agriculture, although the Secretary of State Women's Program was also a contributor.

Concerns about economic survival are constant in the organization, just as they are to individual farmers. As Florence Trautman ably put it, "To do a proper job you need money . . . just like in farming. . . . I'm sure we'll come up with some sort of funding and we'll manage. We're quite versatile: we're farmers."

Young People's Work

After setting up committees on health, education and social welfare, the United Farm Women of Alberta turned next to the important task of organizing young people in their communities. They were concerned about ensuring a membership for the future farm organizations and so the primary focus of young people's work became leadership training and educational activities. The junior branch of the organization, which became known as the Junior UFA, was created in 1919. Mrs. Susan Gunn was named the first young people's convenor and, with funds provided by the UFA, toured the province to organize junior locals. Mrs. Lowe, while president of the UFWA, described it in one of her radio broadcasts as "a fourfold program of development . . . educational, vocational, economic and social. Through conducting meetings, taking part in discussion, the Juniors received training in citizenship and leadership."[2] "Suggestive programs" following the fourfold program of development were produced in the 1930s, similar to the program booklets sent to UFWA locals.

There was not a junior branch affiliated with every UFA or UFWA local, but by 1925, 1000 teenagers were members of locals scattered all over the province. The number of junior locals continued to increase for several decades, so that ten years later the membership total had doubled, to 2000 youths. A convener for young people's work, usually a board member, was charged with organizing and advising the junior branch on behalf of the UFWA. Some of the locals were organized by the Junior president, as he visited UFA locals. Others were organized by directors of the UFWA and UFA in their respective areas. Molly Coupland was a junior member who organized the Wilson junior local and she later became a key figure in the UFWA and FWUA, serving as a director and the handicrafts convener for many years.

The first major project of the Junior UFA, instigated by Irene Parlby in 1919, was the development of Farm Young People's Week at the University of Alberta every June. A member of each junior local was selected to attend the week on the university campus. A fund for travel expenses was set up by the UFA and UFWA and every local contributed to it. The program included lectures, contests, talks by educators and leaders, recreation periods, a business session and lots of socializing. The contests included a scholarship contest, the prize for which was a free term at Olds School of Agriculture; a grain judging contest; and a public speaking contest. There were also two contests for girls only, on "art in the home" and "food for the family," for which cash prizes were awarded. It was a busy and eye-opening week for young people. As Elenore Price, convener for young people's work in 1933, wrote in a bulletin to UFWA members, "No one young or old can ever be the same again after one of these weeks at the university."[3]

The Junior UFA held other contests and activities throughout the year, including a handicrafts contest for locals, an efficiency contest and an essay contest, with the first prize essay published in the UFA paper. A reading course was developed by the Department of Extension at the University of Alberta for the Junior Branch, and books were sent out to the locals to enable their participation in the course. The course formed the basis of a contest among the locals for scholarships at the university. District and, by 1936, provincial conferences brought young people together who did not have an opportunity to attend the university week. Through their affiliation with the Junior UFA, young farm people also had opportunities to attend national and international youth congresses which had programs focused on peace and economic progress.

In 1935 Junior UFA decided it wanted to affiliate with the Cooperative Commonwealth Youth Movement, and resolutions were passed at the UFA convention in January 1936 to enable them to do so, with assistance from the UFA board. The Junior UFA program survived the amalgamation and reorganization of the parent organization that took place in 1949. However, by 1959, the program was less popular with young farm people. The farm youth leadership program gained new impetus with the idea of creating a summer leadership camp at Goldeye Lake.

Farm women were also instrumental in the development of 4-H clubs for young people in Alberta. They served as leaders and instructors in the clubs, and lobbied for government support of 4-H programs.

Goldeye Lake Camp

Goldeye Lake camp is another example of farm women and men's commitment to the development of leadership and citizenship in young people. Goldeye Lake camp was the inspiration of a few individuals in the Farmers' Union and the Farm Women's Union. The original concept of a leadership camp for young people was endorsed at the 1958 annual convention of the FUA and FWUA. An action committee was formed and a fund-raising program initiated. The original funding goal was 78 000 dollars. By 1959, the Goldeye Lake site had been chosen and a sod turning ceremony took place that summer. Situated on lakeshore property leased from the Government of Alberta, Goldeye Lake camp is a beautiful natural site nine kilometres west of Nordegg. Building the camp was a huge job. Lena Haywood recalled that her late husband, Morley Bradley, and Hazel Braithwaite were the first camp leaders when the first leadership pilot camp was held in 1961. "The first camp was held outside . . . there was no building where they had their meeting, [when] one of the first groups of young people came they sat on the logs and they held their leadership camp. They had two log cabins, and they built the washroom." A camp committee report written by the acting chair Gerald Schuler in 1961 identified the physical aspects of the site as "a cook shelter, a dormitory cabin and a wash house; recreation area to be cleared, three dormitory cabins started and to be completed by the end of July; well to be completed by end of July."[4] He also wrote that a personal donation had been received for a barbecue to be built. Fund-raising for the camp was not as successful as founders had hoped and in 1961 a publicity director, Joe Clark, was hired to "revitalize interest, purpose and responsibility." A "widespread public speaking circuit" was planned for the summer of 1961 as well as radio and television appearances, displays and news releases.[5]

Farm women have been heavily involved in Goldeye Lake, building latrines and other facilities, raising operational and construction money, and working as camp counsellors and cooks. Each farm women's local usually sponsored a young person from their area to attend a camp in the summertime. The farm women's and men's organizations jointly carried the expenses and management responsibilities of the Goldeye Lake centre until 1978, when they decided to turn these over to the Rural Education and Development Association (REDA). The Association formed a Goldeye Lake Foundation Board to raise funds for the camp

Goldeye Lake camp.

and to manage it, and a program committee to assist REDA in the development of leadership and educational programs at Goldeye. A Women of Unifarm board member sits on each of these boards.

Every summer, Women of Unifarm sends two of its members as volunteer counsellors to each camp session. Many women have served in this capacity over the years, and most speak highly of the experience and the young people with whom they worked. Figure 5, Appendix B, is a report by Olga Manderson after her leadership camp experience. Helen Murray shared her feelings about Goldeye: "I've been there four times and it's one of the best experiences of my life. . . . It's a beautiful spot and I like working with the young people. . . . I think Goldeye Camp is outstanding. . . . And I don't think anybody ever came home from there without a little training." Many farm women are pleased that their own children had opportunities to attend leadership camps at Goldeye. Campers were usually requested to present a report of their camp experiences to the local that sponsored them.

One of the innovative programs introduced at Goldeye in 1962 by Hazel Braithwaite was citizenship camps integrating Indian and non-Indian youths. A small group of native youths were brought from Fort Chipewyan to join other youths from all over the province. From all accounts, the experiment was a success, and facilitated a great deal of cross-cultural learning. Florence Scissons remembered how careful the organization was not to advertise the cross-cultural camps, so that the news media would not know about it and misrepresent it. "So we kept a lid on it, because it wasn't a publicity thing. We did it because we wanted

Leadership group session at Goldeye Lake camp.

the whites and the Indians to have this joint experience. It was very useful. We had that for several years."

In 1965, the FWUA broadened the perspective of the camp to include young people from all racial backgrounds. In 1966 the purpose of the citizenship camp was to focus on leadership training specifically. After offering the citizenship camp for five summers, from 1962 to 1966, the organization, under the leadership of Louise Johnston, approached the Alberta Department of Youth to develop an evaluation study of the citizenship camp in 1967. An independent and confidential study was carried out by a University of Alberta professor and several students, sponsored by the Department of Youth on behalf of the FWUA. Although the final report found that the camp they studied in 1967 did not meet its specific goals, it nevertheless recommended that citizenship camps continue to be held at Goldeye, as "this type of camp is so worthwhile."[6]

Goldeye Lake centre has broadened its mandate to include training programs for adults as well as youths. It has been favored with the dedication of many farm women over the years, both in raising funds and providing active leadership. Mabel Barker was honored at a dinner in Calgary in the late 1980s, which was attended by 350 people. At her request the dinner was a fund raiser, with all the proceeds going to Goldeye. Betty Pedersen remained active on the Goldeye Foundation Board long after she had to curtail her involvement in Women of Unifarm. Goldeye has a special place in the hearts of many Alberta farm women.

Associated Country Women of the World

The Associated Country Women of the World (ACWW) was formed in Stockholm in 1933. It was initially discussed at a meeting of rural women's groups from 23 countries, which was held in 1929 in conjunction with the International Council of Women in London. An interim meeting was held in 1930 in Vienna, again at the International Council of Women, where a Liaison Committee of Rural Women's and Homemakers Associations was formed, in preparation for the birth of the ACWW in 1933. The organizing principle of the ACWW was to create a permanent link between all the various rural women's groups. Each of its member organizations were to maintain their own autonomy and continue their own work. The ACWW was not to be a new organization of rural women, but a forum in which all the member groups would send representation and participate on an equal basis. Its main aims are to help further international friendship and understanding, to create helpful relations and to raise the standards of living in rural areas everywhere. A report by May Huddlestun of the FWUA, following her attendance at the 1968 ACWW triennial conference in East Lansing, Michigan, stated:

> The ACWW works for improved living conditions and better homes, and
> encourages women to take their place in community life . . . It provides
> scholarships to help improve the standard of living in developing countries.
> It keeps the country woman's point of view before United Nations, where it
> has consultative status on the Food and Agriculture Organization, UN-
> ESCO [United Nations Educational, Scientific and Cultural Organization]
> and UNICEF [United Nations Children's Fund].[7]

In addition the ACWW has consultative status with the Economic and
Social Council of the United Nations, which is responsible for all aspects
of the United Nations, including social development and the Status of
Women. Specific activities of the ACWW have included work on nutrition, child care, family planning and illiteracy. All are focused on a better
life for rural women throughout the world. The philosophy of ACWW is
that the world depends on cooperation, not conflict, and that the heart of
a woman should know no barriers of race, creed or nationality.

ACWW has grown to a large network of over 9 000 000 women in
more than 60 countries. It supports programs organized locally by members and promotes basic education to fight hunger and disease. It also
trains local leaders to spread knowledge gained through sponsored
programs in nutrition and education. It speaks for rural women and
homemakers at all United Nations meetings, and administers grants
given by the United Nations for field work.

The ACWW has provided an opportunity for Alberta farm women to
take part in world affairs and in so doing to broaden their perspectives
and their knowledge. The organization joined ACWW in 1951. Each farm
woman who joined Farm Women' s Union, and later Women of Unifarm,
automatically became a member of the ACWW. As a member organization, Alberta farm women were entitled to send delegates and visitors to
each triennial conference. Depending on the location and guidelines of
the conferences, the organization sent delegations varying from between
two and 16 women. ACWW conferences were of enormous size, varying
from 1500 to over 3000 participants, depending on the venue. ACWW is
the largest women's conference in the world. Each delegate brought back
reports on the work of the Association and a new understanding of the
situation of other rural women in many parts of the world. Sometimes
lasting friendships with women in other countries were forged, and
provided years of personal enjoyment and education through the exchange of letters and sometimes also visits.

The main source of operating funds for the ACWW, supporting 75
percent of its work, has been "Pennies for Friendship." At every Alberta
farm women's gathering, at the local level up to the provincial convention, time is set aside to have a march for friendship and gather the
pennies (in later years, loonies) that go to supporting the organization's
work around the world. Over the years thousands of dollars have been
sent from Alberta to the world headquarters from these coin collections.

Japanese Business Women's Tour/Fort Saskatchewan Women of Unifarm—the two groups met in 1991 in Edmonton to share ideas about business and family life via photos and two interpreters. (photo courtesy Corinne Thompson)

As May Huddlestun of the FWUA wrote in 1968, "These developing countries are not asking the developed countries for charity or handouts, but they want help in setting up self help programs to teach them such things as nutrition, child care, homemaking, farming and above all how to read and write and it is through our pennies for friendship that we can help them."[8]

In 1959 the ACWW set up the Lady Aberdeen scholarship, named after one of its founders. The scholarship is for leaders in the field of home economics, rural community welfare and training in citizenship. Alberta farm women have contributed regularly to the maintenance of this scholarship fund. May Huddlestun, an FWUA delegate to the 16th triennial, mentioned in her convention report that "We met some of the Lady Aberdeen scholarship winners who now hold important positions in their governments."[9] In 1980, the 16th triennial conference called on all women to support the International Women's Year "Plan of Action" to strengthen international peace and disarmament and the development of friendly relations between nations. Another recommendation at the same conference was about breast cancer. It urged members to activate women as well as health authorities in their countries to promote the earliest detection of breast cancer. Other concerns of ACWW expressed at this and subsequent conferences were the use of infant milk powders; the violence in entertainment in press, television and film; the exploitation of developing countries by the export by other countries of inferior goods; and a resolution requesting that the United Nations consider pornography involving children a breach of human rights. Other resolutions in the eighties focused on the need for government action to reduce acid rain, the reduction of nuclear arms, the control of pornography and the elimination of discrimination between sexes, races and nations. A resolution also stated that all member societies should emphasize the equality of women within the home, place of employment, local community and national government. These are just a few examples of the broad range of issues considered by the ACWW in its conference programs.

Jacqueline Galloway attended the ACWW triennial conference in Kansas City in 1989. She remarked:

> *Sitting in a room with 1300 women of every color, every culture, every language and hearing about their lives, it's just wonderful. . . . You know we can bring a resolution from the Women of Unifarm, Fort Saskatchewan local, to the district, to the provincial organization and have that resolution go on to the international conference and have it presented at the United Nations. It's an incredible pipeline. ACWW is very action oriented. It's not sitting around talking, they fund projects that directly improve the lives of women in developing countries.*

The 17th triennial conference of the ACWW was hosted by Canada, in Vancouver. Alberta farm women became actively involved in the host activities of the conference. Committees focusing on finance, volunteers, pre- and post-convention tours, displays and exhibits, a choir, Canada Day, public relations, a craft sales table and a stamp collection were set up by Women of Unifarm. In addition, under the leadership of Vera Rude, 2200 wheat favors, designed by Katharine Russell, were manufactured from home grown wheat by Women of Unifarm members to place in the conference mailboxes of foreign visitors. They were well received. Because of the proximity of the conference site, many Alberta farm women were able to take part, as members of a choir, as volunteers, as delegates and as visitors. Many women recalled the experience as one of the highlights of their lives. Alberta farm families also enjoyed hosting foreign delegates on tours through central Alberta.

There is no doubt that the lives and the work of Alberta farm women have been enhanced through their affiliation with the work of the ACWW. Many participants in the ACWW conferences over the years described it as an educational experience they would never forget. An education in international citizenship, an opportunity to serve humanity beyond the country's borders, and a chance to refocus on their own work, within Alberta, with a fresh perspective, are just some of the positive effects of this important connection. Inga Marr wrote, after attending the ACWW triennial conference in Lansing, Michigan, in 1968:

I felt the world growing smaller and smaller as the days progressed, as women from all parts of the world mingled together, communicating ideas and discussing common problems. I realized as never before what power women have, what a tremendous responsibility is ours, as mothers, as educators of the future generation. Women have the greatest influence in attitudes of our youth, but I feel that we in this land do not always fulfil our potential. Women from the developing countries seem to feel this tremendous responsibility and be willing to accept it.[10]

As Jean Ross pointed out after her participation in the 1983 ACWW conference in Vancouver, "It helps us to realize that we are not the only frog in the puddle . . . It's good to know these things."[11]

International Projects and World Peace

Alberta farm women were involved in other international projects of diverse types. Foster children in developing countries were sponsored by locals and also by the board on behalf of the membership. For many years, farm women cooperated with Operation Eyesight, gathering eye glasses to be sent to citizens of developing countries. In 1951, the Farm Women's Union participated in a "march for diapers" for the infant children of Greece. The organization was also a regular supporter of the work of the Unitarian Service Committee and of Amnesty International.

Prairie farm women are well known for their keen interest in working toward world peace. Most of their peace activism took place between the two World Wars. The farm organization was founded in the middle of the first World War. With their own grief and horror fresh in their memories, organized farm women embraced a partnership with the Women's International League for Peace and Freedom without hesitation. Individual locals as well as the board of directors affiliated with the League. Both the UFWA and the Women's section of the Canadian Council of Agriculture set up committees on peace and arbitration, to study the issues, make recommendations and coordinate actions with the League. The study program of the locals had a focus on international affairs, and in 1921, the topic was the relationship between food and peace. As Mrs. Lowe described the rationale in later years, " We know that all is not well with the world, nor can any section of the country prosper while poverty and starvation stalk any portion of it. We feel that empty stomachs do not provide a foundation for any sort of peaceful attitude on the part of nations."[12]

In 1932 Alberta farm women participated in a nation wide petition for disarmament that was circulated and signed by half a million Canadians and shipped to the League of Nations disarmament convention, along with petitions from many other countries. Alberta farm women traversed the province obtaining signatures on the petition at every farm house that would welcome them. In recognition of the farm women's contributions to the peace effort, Irene Parlby was asked to represent the women of Canada as a Canadian official at the League of Nations conference in Geneva in 1932. The concern for world peace escalated in the '30s as Canadians watched events unfold in Europe. As Mrs. Price described it in 1934, "we have been appalled again and again at the near approach of an outbreak of war."[13] As president of the UFWA, she devoted most of her 1934 convention address to the UFA to the analysis of economic conditions and political events that threatened the security of all nations. Farm women's peace activism turned to support for the war effort as 1939 brought Canada once again into international conflict. Many Alberta farm women later joined the women's peace and disarmament organization, the "Voice of Women" (VOW), which was organized in 1960. Some remain active in VOW to this day.

Alberta Farm Women's Network

Women of Unifarm has participated as a member of the Alberta Farm Women's Network since its inception in 1984. The Network grew from a nucleus of five farm women, four of them members of Women of Unifarm, who were taking part in a workshop that focused on bringing information and control together in order to make changes in women's lives. The five women decided that what they wanted was a network "where all farm women and organizations could come together and work for needed changes that affected their lives and livelihood."[14] Another goal was to attempt to avoid overlap in working on common issues and in accessing government funding. Joyce Templeton, who was one of the nucleus group members, pointed out with pride that of all the women's groups represented at the workshop, only farm women took action to develop a provincial network.

Two months after the workshop, in November 1984, the nucleus group invited others to join them at the Women of Unifarm convention to "set up goals and committees for the First Alberta Farm Women's Conference."[15] The conference was held in Red Deer in January 1986. At the national level, the federal government had sponsored the second national conference of farm women in November 1985. Jean Buit, President of Women of Unifarm, had acted as the Alberta contact with the national planning group. The first national conference of farm women had been held in Winnipeg in 1980, and had focused on the image of farm women, their economic contributions and their access to credit. According to Mary Newton, there was a strong feeling at the second national conference in 1985 that there was a need for a national forum on a more regular basis than a major conference every five years. In the spring of 1986 representatives of each province met in Ottawa and a firm commitment was made by all in attendance to return to their respective provinces and start the development of provincial networks. Mary Newton assumed the role of provincial coordinator at the request of Jean Buit, who had retired. In June 1986 representatives from the National Farmers' Union, the Christian Farmers Federation, Women of Unifarm, Women in Support of Agriculture and the Alberta Women's Institutes met at the Red Deer conference and decided to form the Alberta Farm Women's Network. Representation was set at two members from each member organization, and two members at large.

Plans for the Network included an information brochure and a quarterly newsletter. The same group met again in late June to share ideas and to develop plans and a budget. The first newsletter was produced in the fall of 1986. Another national conference was planned for November 1987 in Saskatoon, and delegates from Alberta learned that Alberta was the only province that had contacted all major farm groups and made an effort to form a province-wide network. Another meeting was soon called by Agriculture Canada, with the intent of organizing a Canadian Farm Women's Network. As Mary Newton described it, "pressure seemed to be building from the top down."[16] Agriculture Canada and the Secretary of State were pressing for the development of a national farm women's organization, and a discussion paper was developed to that effect. The paper was followed by a proposed constitution for the new national organization. Some provinces, including Alberta, became concerned that the networking concept was lost, and instead the federal government was creating another farm organization for women. It appeared that it would be an organization whose mandate and procedures would be determined by the federal government, not by farm women. It also seemed to serve the government's interest in simplifying the consultation process, and controlling the farmers' lobby. Alberta chose to become an associate member of the new national network rather than a full voting member because of concerns about the nature and purpose of the national organization.

At the provincial level, the Network continued to thrive, offering a biennial conference for farm women to get together to discuss common concerns and issues. These issues included stress and responses to it, marketing, environmental concerns, linkages between food producers and consumers, and networking with urban women and women's groups. The Network also sponsors the Alberta Farm Woman of the Year Award, which recognizes an outstanding farm woman who has contributed both to the industry and to her community. Members of the Network find that it has removed the barriers that previously existed between farm women's organizations in Alberta. It provides an opportunity to listen to each other, to develop a kind of consensus while still maintaining some diversity, and to enable linkages between people and issues. As Margaret Blanchard said, "The original idea of having an exchange of information and avoiding duplication is the one thing I think is important . . . I wouldn't like to see it become a voice for all women farmers in Alberta because I don't think you could get that much unanimity. I think we need the diverse opinions."

Mary Newton likes the results the Network is achieving.

We've had pretty good communication with all the other farm groups and I think that's really important . . . It could really grow a lot and I hope it does because it serves a real purpose. . . . We have developed good communication among the groups that are working together and a real warm feeling among groups that were always traditionally said to be antagonistic and we've overcome that . . . we didn't want to form another organization and we've continued to stay away from doing so.

In 1991, the Network launched a talent bank directory project to encourage the participation of farm women on provincial and civic boards and commissions. They communicate effectively with their growing membership through a quarterly newsletter, and continue to organize, with the sponsorship of the UFA, the Farm Woman of the Year Award, which is presented at their biennial conference.

Preserving the Past

Alberta farm women appreciate their heritage as women and as organized farmers. Many individuals and locals over the years put considerable effort into preserving the histories of their communities, their families and their organizations though the development of scrapbooks and history books, and the establishment of local museums. It was the Farm Women's Union of Alberta that lobbied the Alberta government to establish a provincial archives to hold the documentary heritage of the province. Members were encouraged in their historical preservation activities by scrapbook contests often featuring the history of their locals, which were judged at annual conventions. Members of locals that won these contests are proud of their achievement. The scrapbooks represent a collaborative and cooperative endeavor that reflects a respect for the things that women value most about their lives. Several locals also entered and, in two cases, won history book competitions for community and regional histories sponsored by the Alberta government.

Conservation of buildings and of land were also a concern to farm women. In a 1972 brief to the Environment Conservation Authority on Conservation of Historical and Archaeological Resources in Alberta, Women of Unifarm stressed the importance of the preservation of historic buildings in Calgary; the preservation of sites in the country that were good for recreation areas; and a closer guardianship of the badlands of the Red Deer River valley. In addition they made specific suggestions for the development and maintenance of heritage sites and services in the province.

Farm women speak of two special kinds of heritage that they hold very dear, and which truly are part of them and their commitment to their lifestyles. One is their heritage of their ties to the land; the other their heritage of service to others. Both of these very profound feelings have

Display set up at the Chestermere Complex for Pioneer Days 1992. Note the cookbooks. (photo courtesy Dorothy E. Clayton)

roots in the generations of women who came before them and settled the prairie, and passed these values on to daughters and sons. For those who struggled to stay and make a living on a prairie farm, the land demanded everything one had, and offered rewards and punishments at random. Women grew to have intense feelings about the land, either love or hate and sometimes both feelings at once. Women related to its beauty, its mystery, its fragility and its power. But ultimately, the land, their land in particular, had special meaning because it was something they fought to keep and that they cared for on a daily basis. Farm women of every generation expressed this important relationship to the land, and the sense of history and continuity they feel with the generations of men and women before them as they live on it and work on it.

The first generation of farm women reached out to each other in times of need, and then as organizations were formed, they carried their neighborly helpfulness and concern into the organizations with them. What was born of necessity matured in ways that generations of farm women would emulate. This heritage of service is described by many women of farm backgrounds as a family tradition, something they learned watching grandparents and parents. It was simply part of life for many families. As Verna Kett described it in her family:

My grandmother was very involved. My mother was very involved and so it sort of grows on you and you don't realize that for a long time. My grandmother was very outspoken, a pioneer from Wales and she worked on rights and that kind of thing . . . I think it is sort of inbred in you that you want to hold that tie. Do the best you can through whatever talents you've been allotted.

Like many farm women, Verna grew up in the organization. She was in the youth program before she joined the women's organization. Other farm women spoke about watching their mothers as they did their organizational work, and remembering the significance of that work to the community and their mothers. Many women expressed the wish to see a history written for their mothers' pleasure, because as daughters they share and understand the memories their mothers cherish of their years of fellowship and service.

Facing the Future

For many years the organizations of farmers were able to serve their members well and in turn be well supported by their membership, simply by responding to the current issues and challenges their members faced. Men and women considered their organizations effective mechanisms to represent them and their interests, and to bring about change. The history of the farm women's organizations as they reconstituted themselves over the years demonstrates a willingness to adapt to changing times and a desire to tackle most concerns brought to local or provincial meetings. But in 1970 their role began to change as farming changed to become more specialized, and as commodity groups, representing those special interests of farmers, became stronger and viewed as an effective way to influence policy-makers. Other organizations, task forces and committees were also more frequently involved in planning and implementing policies and programs that affected the quality of life in rural communities.

At the same time that these shifts were taking place, the nature of the family farm was changing. More and more women were involved in off-farm employment as their income became essential to maintaining the family farm. In some cases, men were employed in off-farm work, and farming became a second occupation, sometimes leaving the women to manage the farm mostly on their own. As family time and resources became stretched over more activities, farm organizations found their membership rolls decreasing. Young farm women simply did not have the time for meetings and other organizational activities.

The changes in membership brought about some real uncertainty among the farm women about their role as a provincial organization. Some of this uncertainty was channelled into a concern for the image of farm women in society. This focus was supported in part by the interest of the federal government in this area, as the first national farm women's conference was launched in 1980. A form of identity seeking and affirmation, this activity helped modern farm women come to terms with some of the changes in the farm community and with their place in it. But that process is ongoing. Some older farm women do not understand the new and different pressures that young farm women face, and criticize their focus on their careers, their children and their children's activities. Young farm women speak about a need for an organization that addresses their issues more directly and consistently, particularly the financial pressures on young farm families resulting from the economic crises of the 1980s.

Florence Trautman said she can understand how young farm women, busy raising young children, do not have time for the organization, because she felt the same way once. "I remember when we first got married the local tried to get us involved and we were busy raising kids, and I said, 'I haven't got time for this,' and now I'm having the same problem trying to get somebody else involved as they had trying to get me involved." She also recognized that the needs of young farm women are different today from the women of previous generations:

> *Their needs for Women of Unifarm . . . at that time they were working side by side with their husband all day long. With the kids and everything they needed to get away. They needed to get away with somebody that had the same problems they did. Where I'm finding now in most cases a lot of the farm women are working off the farm. They need to get together with their husbands in the evening. And maybe go together to meetings if they possibly could.*

Betty Pedersen pointed to the changes at the family farm level in the partnership between men and women, with women becoming equal partners in both the work of the farm and the farm decision-making. In her day, she claimed, the partnership could be described as, "You do your role and I do my role and you're still subservient to me." Although she sees the farm partnerships changing, she feels it is not reflected in the farm organization. "I'd like to have some changes in the chauvinistic attitudes in the men's organization, but that is beyond anything I could do. A changing of attitudes, and those change slowly." Verna Kett, who served as an alternate director for many years, then later served as a director, vice-president and president, claimed that she found, sitting on the Unifarm board, that "Some men are very chauvinistic and there's no way that a woman could have an opinion that's okay." She suggested that some farm men are still afraid of a strong farm women's organization.

Some of the farm women who were interviewed suggested that farmers in general have a hard time working together because they are too independent by nature. Women who were selling memberships for years, and encouraging neighbors to become involved, gave up this activity, "stumped" and "stunned" by the attitudes they encountered. The belief in self-reliance as a virtue seems to counteract the willingness to work together with other farmers. As Helen Murray put it, "Because they think they are independent, you see, that there is nobody behind them looking after them. I think in this old world we have to work together and I think women can possibly do it a little better than men. And if a husband doesn't want to join then often the woman isn't going to be in on it either."

Margaret Blanchard, president from 1987 to 1989, reflected on what these realities mean for the future of the organization:

> [we should be looking] to see what can be done to help rural areas keep vital . . . Sometimes an idea comes from a very small gathering or maybe two or three people talking together that can be a really worthwhile idea when it's expanded and I think that's something that Women of Unifarm still have to do is to get those ideas from the people in the country . . . I don't think an organization is worth anything if it's only top heavy, I mean all the ideas come from the top . . . I think we're going to have to find some way to get ideas from the younger people. . . . I think we have to find some way to get that flow of information from them of ideas too. . . . 'cause they're the ones that are having to cope with the problems . . . I think somehow we have to get that energy from them. . . . we have to have people who are looking at the way it is now.

A recurring concern raised in interviews with farm women as they face the future is a need to address particular issues in a more comprehensive way. The most common issues mentioned were rural child care; environmental issues, including the use of chemicals; and the economics of farming. Some established women farmers also suggested their focus has changed to the issues that are affecting the survival of their occupation, and they want farm organizations to work more diligently on these issues.

The challenge of renewing and strengthening the organization with the participation of some of the young farm women was a concern widely shared by members of the organization as they contemplated its future. Elizabeth Olsen, president in 1990 and 1991, responded to a question about the future direction of the organization:

> I see us working more closely with Unifarm in liaison with them in a better way than we have. . . . the challenge is reaching out and getting the younger women, maybe not on a rigid or regular basis, maybe on a flexible basis. . . . They are interested but they have to prioritize their time and we have to work with that and hopefully come up with an answer that would keep them at least in contact with us.

Jacqueline Galloway suggested that it is critical for locals to identify what is important to younger members and to provide that. She sees the need for locals to design themselves according to what the members want to do. Colleen Casey Cyr agreed with this approach. She felt the local in which she was a member did not really help her as a young farm woman, and that the organization could do more to assist young farmers who are in economic crises, at risk of losing their farms.

Some women still envision a day when, as Elizabeth Olsen stated it, "we'll be so independent, we'll be running in Unifarm." Still, Elizabeth feels that day may never come and that "we'll be here as long as we're needed and as long as we provide a voice for those who need us. And if it gets to the stage that the younger women coming up feel very comfortable working within the main organization, then they probably won't need us. But that remains to be seen."

Jacqueline Galloway believes that Women of Unifarm has an important role at the local level, as a support group and as an opportunity for personal development, but "that men and women must talk to each other and educate each other on all the issues." She feels it is time that quality of life issues for rural Albertans "should be dealt with under the general farm organization" and that both male and female perspectives are needed to make good balanced decisions on all issues. Although she thinks there is nothing to stop a woman from getting involved in Unifarm, the "very existence of the two groups perpetuates an inequality; it's segregating, defining certain things as women's issues and certain things as men's." Mary Newton thinks that while women's perspectives are vitally important, Women of Unifarm has at times sat back and let the men do it. Women of Unifarm, she felt, should be working more on farm economic issues.

Like other farm women, Jacqueline Galloway feels that farm organizations are in a critical period of transition, questioning whether to stay with the traditional ways of doing things or to try new approaches. She suggests that in another ten years, "we're going to see so many women and men working together that it won't be an issue any more." Farm women are honest about the changes in their organization, and vocal about the isolation they feel from the decision makers at the board table. The task for the future will be to shape new forms of organizations that serve members who face new and different challenges as farmers.

Though they often speak modestly of their efforts and achievements, farm women tackle issues with intelligence, a diversity of talents and an indomitable spirit. They have survived many transitions and crises, both in the organization and on their own farms. Their belief in the value of farming as a family enterprise has always been their mainstay. Alberta farm women have used the skills and knowledge they gained through their organizational involvement to respond to the needs of the community, to provide effective leadership and to reach out to the world beyond the province's borders. With compassion and conviction, they live as active citizens.

Conclusion

This history was written to identify the contributions of farm women to the life of the province and to farming. It recognizes their significant role in the farm movement. It acknowledges the heritage of service and of struggle that farm women have passed from one generation to the next. It is, as one farm woman described it, the chance to stop and enjoy the view as you climb the mountain.

It is important to acknowledge that the climb up the mountain has not been an easy one. It has also been one largely excluded in the popular accounts of Alberta's history. Without the dedication, hard work and the vision that farm women brought to their community and organizational work, many fundamental services to Albertans would not exist. From the first generation of homesteaders to today, farm women invested themselves in the betterment of the society in which they live, pushing for reform, for resources and for reasonable decision making that put the welfare of people first. Most essential services, and also many innovative social welfare, health, recreation and education programs, were established because farm women dreamed of them, fought for them and maintained them. The commitment they shared in the family enterprise extended to the larger community beyond the farm gate. Meeting this commitment came at great cost, both in economic and human terms. It also had its rewards, in the friendships that were developed and enjoyed, the personal growth women experienced and the sense of accomplishment they felt when they knew their work had made a difference.

Organized farm women knew politics and social reform from the bottom up. They knew the human side of every issue, the impact it may have on a neighbor, a friend, a loved one. They lived their politics daily, as it shaped their lives and their livelihoods. They understood the power of organization to influence the political process, and they learned, through their organizations, how to use it. They were aware of the gender politics within their community and their organizations, and how it influenced priorities and policy outcomes. They understood the relationship between farm men and farm women in a variety of ways, but most were keenly aware that women do more than their share of both the work and the compromising in farm partnerships and in farm organizations. Alberta farm women, through their vision of community and of farming, and their commitment to work hard to achieve that vision, have created the caring and progressive society Alberta is today. All Albertans, urban and rural, are in their debt.

Epilogue

After this book was written, and before it went to print, some major changes took place in the Unifarm/Women of Unifarm organizations. After many years of struggle to curb the declining membership and to redefine its mandate, Unifarm voted at its January 1996 convention to disband Unifarm and set up a new direct-funded producer organization called Wild Rose Agricultural Producers. Government funding and the funds from commodity association members would no longer sustain the new organization, which chose to return to its historical roots and be a "grass-roots" collective voice for individual producers. With its new look came some confusion about its relationship to Women of Unifarm and some controversy about the family and individual membership privileges in the new organization.

In response to the changes in Unifarm and with the understanding that there were no "family" memberships in Wild Rose, participants in the Women of Unifarm annual convention in June 1996 discussed the future of Women of Unifarm. A group of 40 women met at the convention in Lacombe and in the heat of discussion, several issues emerged. One was the reticence to change the organization's name once again to affiliate with the new Wild Rose Agricultural Producers. Another was the anger women felt that their share of membership fees, which amounted to $35 000, had been withheld from them by Unifarm as the parent organization struggled financially and folded. And as Ron Leonhardt, President of Wild Rose, told the women's convention, there was a lot of concern about the viability of the new organization, because of financial problems and low response to membership drives.

Elizabeth Olsen moved that the women's organization be dissolved in consideration of the age and number of its members: there were simply too few to sustain it and too many were in their senior years. Other members voiced their concerns that if the women continue, they had to commit to making it work. Judy Pimm, a younger member, felt it was premature to disband without giving the organization one last try. Finally the vote on the motion came and it was defeated. The women at the 1996 convention of Women of Unifarm decided to maintain their organization, keep their name and declare their independence. They would not affiliate with Wild Rose, and they would continue to try to represent the interests of Alberta farm women. New president of Women of Unifarm, Florence Trautman declared, "We have given birth to a new organization . . . we have faith that what we're doing is beneficial to all farm women." A committee was struck to work on a new constitution and a new initiative in support of a hearing loss project was launched. Still, the Women of Unifarm members gathered at Lacombe in June 1996 recognize that their members are disappearing and not being replaced, that the money required to support the organization is limited and that many previous sources of funding are no longer available to them. They no longer represent a wide diversity of farm women across the province and the amiable partnership with the larger organization has dissolved into acrimony. The organization has matured in the process of negotiating the changes that the 1990s have brought. It will be interesting to watch the organization as it meets the challenges that face it in the late 1990s.

Notes - Chapter One

1. K. Norrie and D. Owram, *A history of the Canadian economy* (Toronto: Harcourt Brace Jovanovich, 1991), p. 325; Government of Canada, *Census of the prairie provinces*, 1926, p. ix.
2. N. F. Priestly and E. B. Swindlehurst, *Furrows, faith and fellowship* (Edmonton: Co-op Press, 1967), pp. 42-43.
3. *Working hints for local unions of The United Farm Women of Alberta*, p. 1.
4. I. Parlby, "A while ago and today," *Canadian Magazine*, 1928.
5. I. Parlby, "A tale of a little club," 1915.
6. Minutes of Edgerton local, June 1913.
7. Minutes of Edgerton local, February 1913.
8. Correspondence of P. P. Woodbridge to Irma Stocking of Saskatchewan Women Grain Growers, Delisle, Saskatchewan, January 4, 1915. Answered by Violet McNaughton, January 11, 1915. Violet McNaughton Papers, Saskatchewan Archives Board.
9. *The U.F.W.A.: The organization for Alberta farm women*, p. 1.
10. Resolution to the 1912 convention of the UFA by the president, W. J. Tregillus, "be it resolved that we believe the wives and daughters of our farmers should organize locally and provincially." *UFA annual report*, 1912.
11. Correspondence of P. P. Woodbridge to Irma Stocking, January 4, 1915. Violet McNaughton Papers, Saskatchewan Archives Board.
12. Ibid.
13. L. Barritt, *The United Farm Women of Alberta*, 1934.
14. I. Parlby, taped interview, 1958, Provincial Archives of Alberta, 67.218.
15. E. Carter, *Thirty years of progress*, 1944.
16. I. Parlby, "A tale of a little club," 1915.
17. I. Parlby, article on farm women's clubs, *Grain Growers Guide,* April 14, 1915.
18. Mrs. Spencer, memoirs, 1952.
19. I. Parlby, taped interview, 1958, Provincial Archives of Alberta, 67.218.
20. These new locals included Calgary, Blackie, High River, Cayley, Nanton, Macleod, Craigmyle, Delai and Stettler.
21. H. Shaw, "Women of Unifarm," *Carstairs Community Press*, January 14, 1981, p. 7.
22. I. Parlby, report of the UFWA President to the UFA, *Annual Report*, 1917, p. 33.
23. Ibid., p. 31.

24. I. Parlby, report of the UFWA President to the UFWA, *Annual Report,* 1917, p. 279.
25. I. Parlby, "Farm women of Alberta," Letter to the *Grain Growers Guide,* August 9, 1916, p. 23.
26. L. Barritt, *The United Farm Women of Alberta,* 1934.

Notes - Chapter Two

1. Pamphlet, The UFWA: The Organization of Alberta Farm Women, n.d., p. 3.
2. Mrs. J. B. Kidd, Report to UFWA Convention, 1922.
3. Irene Parlby, Speech to UFWA Convention, 1934.
4. Pamphlet, The UFWA: The Organization of Alberta Farm Women, n.d., p. 13.
5. Leona Barritt, 1916, quoted in "The Alberta Farm Women's Movement," a paper presented in February 1986 by Barbara Evans.
6. Hazel Braithwaite papers, n.d.
7. Pamphlet, The UFWA: The Organization of Alberta Farm Women, n.d., p. 4.
8. Nellie Peterson, Interview with Barbara Evans, n.d.
9. Directors' Report, Farmers' Union of Alberta Convention Report, 1951.
10. Myrtle Ward, Interview with Barbara Evans, n.d.
11. Roseleaf Local History, n.d.
12. Region One Conference Report, Women of Unifarm Head office files, 1978.
13. Jacquie Jevne, Women of Unifarm Annual Convention, 1984.
14. Region 3 Spring Conference, April 9, 1986.
15. Ibid.
16. Susan Gunn, Presidential Address, United Farmers of Alberta Convention, 1926.
17. FWUA Correspondence to CAC, 1957.
18. Elsie Seefeldt, Convenor's Report, 1986.
19. Women of Unifarm, Brief to Agriculture and Rural Affairs Caucus Committee, 1989.
20. *Grain Growers Guide,* January 15, 1919, p. 47.
21. WSCCA Secretary's Report, 1921, Saskatchewan Archives Board A1 E11.
22. Hazel Braithwaite papers, 1964.
23. Women of Unifarm, Report on Farm Stress, 1977, p. 2.

24. Vera Rude, President's Report, November 26, 1981.

Notes - Chapter Three

1. Winnifred Ross, Winnifred Ross papers, n.d.
2. Violet McNaughton made the presentation to the CNATN special committee in 1916 on behalf of prairie farm women. She proposed a partnership model for maternity care, where both a trained midwife and a nurse would work together in a district. The midwife would handle the routine cases and she would call the nurse to assist with more difficult or complicated births.
3. Vera Lowe, FWUA Convention Report, 1949.
4. FWUA Convention Report, 1961.
5. Ibid.
6. FWUA Health convenor's report, 1951.
7. Women of Uniarm Convention Report, 1977.
8. Delegate's report on annual convention, 1921.
9. Ibid.
10. I. Parlby, Published Speech: Mental Deficiency, An address delivered by the Hon. Mrs. Parlby before the UFWA, January 1924.
11. Mrs. R. Price, President's Report, UFWA Convention Report, 1931.
12. Correspondence from Honourable J. J. Lymburn, 1932.
13. Report of Region 7 conference, 1982.
14. UFWA Brief to provincial cabinet, 1919.
15. Winnifred Ross, Education Bulletin, 1932.
16. Margaret Richardson, memoirs, n.d.
17. FWUA Convention Report, 1961.
18. Ibid.
19. Grandmeadow Local history, n.d.
20. Mrs. R. Price, President's Report, UFWA Convention Report, 1935.
21. Mary Bentley, Social Welfare Bulletin, UFWA, 1929.
22. Conference Report, Region 12 Women of Uniarm, 1977.
23. Conference Report, Region 7 Women of Uniarm, 1978.

Notes - Chapter Four

1. Pamphlet, The UFWA: The Organization for Alberta Farm Women, n.d., p. 5-6.
2. UFWA Program, 1948.

3. Minutes of the Ridgewood Local, 1946.
4. Ibid.
5. Minutes of the Horn Hill Local, 1946.
6. History of the Ridgewood Local, n.d.
7. Minutes of the Ridewood Local, 1942.
8. History of the Ridgewood Local, n.d.
9. Mrs. Banner, Region 11 Conference Report, 1941.

Notes - Chapter Five

1. Lucille MacRae, Report of UFWA Secretary, Annual Convention Report, 1926, p. 76.
2. Vera Lowe, President, UFWA. Broadcast over CFRN, CFGP and CFCN, October 24, 1947.
3. Elenore Price, "The United Farm Women of Alberta Young People's Work," November Bulletin, 1933.
4. Gerald Schuler, *Camp Committee Report*, July 13, 1961.
5. Ibid.
6. *The Goldeye Lake Citizenship Camp. An Evaluative Report.* Research Branch, Department of Youth, Province of Alberta, 1967, p. 87.
7. May Huddlestun, Report on ACWW Conference, 1968.
8. Ibid.
9. Op. cit.
10. Inga Marr, Report on ACWW Conference, 1968.
11. Jean Ross, Report on ACWW Conference, 1983.
12. Vera Lowe, President, UFWA. Broadcast over CFRN, CFGP and CFCN, October 24, 1947.
13. Mrs. R. Price, President's Report, UFWA Annual Convention, 1934.
14. Joyce Templeton, "Early ingredients of the Alberta Farm Women's Network," *Alberta Farm Women's Network Newsletter Vol. 6*, no. 1, p. 2., Feb. 1991.
15. Ibid.
16. Mary Newton, "Networking, Provincially and Nationally." *Alberta Farm Women's Newsletter, Vol. 6.* No. 3, p. 3.

Appendix A

PRESIDENTS OF THE UNITED FARM WOMEN OF ALBERTA
1915 - 1949

Miss Jean Reed (Women's Auxiliary, 1915)
Honourable Irene Parlby
Mrs. Marion Sears
Mrs. Susan Gunn
Mrs. Amy Warr
Mrs. R. Price
Mrs. M. E. Malloy
Mrs. Winnifred Ross
Mrs. Vera Lowe

PRESIDENTS OF FARM WOMEN'S UNION OF ALBERTA
1949 - 1969

Mrs. Vera Lowe
Mrs. Winnifred Ross
Mrs. Ivy Taylor
Mrs. C. T. Armstrong
Mrs. Hazel Braithwaite
Mrs. Louise Johnston
Mrs. Paulina Jasman
Mrs. Elizabeth Pedersen

PRESIDENTS OF WOMEN OF UNIFARM
1970 - 1996

Mrs. Elizabeth Pedersen
Mrs. Inga Marr
Mrs. Leda Jensen
Mrs. Vera Rude
Mrs. Jean Buit
Mrs. Barbara Klymchuk
Mrs. Margaret Blanchard
Mrs. Louise Christiansen
Mrs. Elizabeth Olsen
Mrs. Verna Kett

Appendix B

List of Figures

ALBERTA WOMEN'S WEEK

Alberta Women's Week was held at the Olds Agricultural College, July 19-22, 1982. I found this, my first time in attendance at such a conference, most enjoyable and worthwhile.

The first speaker, Marilyn Ferguson, science writer from Los Angeles, introduced the theme for the week "I make the difference". She said human capacity is greatly underestimated. Each of us has within us a renewable resource of confidence, courage, desire, talent and intuition that just needs to be awakened. Given the right catalyst, the human mind can be transformed, and can in turn bring about a change in society. There is an increasing vision that women will fuel the whole movement towards social transformation. The speaker left a mood of optimism in a time of fear for the future. She closed her presentation very fittingly with a recording of John Denver's "It's in Everyone of Us".

Shirley Meyers, Head of the Home Economics Branch, continued the theme by explaining how we can "make a difference" in our lives and communities by using the many services that are provided by the Home Economics Branch.

The second day opened with a big disappointment and all in attendance could empathize with the organizers when the keynote speaker, Father Larre, failed to appear, cancelling his engagement on very short notice. He was to have carried on the theme "I make the difference —- to my family". However, we were able, in this way, to attend one more "select-a-Session" than planned. The instructors and resource people very obligingly responded on very short notice, to present an extra session.

On Thursday, the speaker Suzanne Trouba, a fashion coordinator from Calgary, continued the week's theme with "I Make the Difference — to Me". She said our appearance is a self-portrait — we make an instant statement about ourselves. Personal appearance has a definite effect upon personal success. She then gave tips on total color coordination and wardrobe planning.

Of the Select-A-Sessions, I attended:

1. Aerobic Fitness and Dance with Avra Teplitsky of Slave Lake.
2. Two-Family Farm - Nadine Vester was unable to attend, but Garry Bradshaw, Regional Farm Economist of Red Deer, dealt with the subject very capably and made it interesting and enjoyable. I felt that this REDA production was excellent and heard many favorable comments on this session.
3. Getting My Two Cents In - Elizabeth Webster, Personal Development Specialist with 4-H, gave advice on public speaking and shared a good film entitled, "Applause".
4. Money Mangement in the 80s - Doug Henderson, broker with Dominion Securities, Calgary, drew a large attendance at all his sessions. He had a lively entertaining way of dealing with a rather heavy topic.

A daily feature which I felt added a great deal to the conference was a short morning and evening service at the College Chapel. Various denominations from the Olds Ministerial Association attended. I was impressed by the way in which they all attempted to integrate their scripture readings and messages with the theme of the conference "I Make the Difference".

Accommodation at the College residence was not fancy, but adequate. The meals at the cafeteria were excellent. I wish to thank Women of Unifarm for the privilege of attending Alberta Women's Week. It was an enjoyable and enriching experience.

September, 1982   Respectfully submitted,
                                            Ursula Delfs, Region 1 Director

POLICY ON INDIAN AFFAIRS

RESOLVED that the Farm Women's Union of Alberta adopt as policy relative to the Indian people, the following:-

1. We respect the right of the Indian people to maintain and develop their own culture.

2. We believe all promises made to the Indians in their original Treaties should be kept.

3. We believe all Canadian Indians should have the same recourse to law as other Canadians, and should have access to a Federal Government appointed lawyer.

4. We believe that all Canadian Indians should evolve to the Canadian citizenship right of the franchise, with no other restrictions or penalties than are applicable to all Canadians.

5. As at this time the majority of the Canadian Indian people are under privileged in comparison to the general Canadian and Alberta education and economic standards and opportunities, we believe this should be remedied as soon as possible, with these programs;

   (a) Full opportunity for elementary, junior high and high school education for all Indian children, and vocational and professional training for each Indian child to the full extent of his capabilities, and to equip her or him for employment within, or outside the reserve, and with at least 25% increase in grants-in-aid to be made to those students at public, high, technical or business schools outside the reserve, to meet the increased costs of tuition, books, board and lodging, and with the Indian Act statement that Indian children be not forced to attend either Protestant or Catholic Schools counter to their faith, adhered to, completely impartially.

   (b) Research and assistance in improving occupational methods and output;

   (c) Monetary and personal assistance, and every encouragement possible in developing native skills and arts;

   (d) citizenship training of vision and sensitivity and appreciation of Indian culture, and leadership training of the elected Treaty Reserve Councils, and Indian community leaders other than of the Treaty Reserves;

6. We believe that the Indian Affairs Branch should give employee preference to, develop vocational opportunities for, and give every encouragement to, Indian young people with special and professional training to return to the Reserves with these advantages in mind;

   (a) Assisting the maintenance of the Indian cultures;

   (b) Raising the standards of life among the Indian people with the insight and appreciation of their needs being better envisioned and served by their own trained young people;

   (c) The dual stimulation presented to Indian Youth; the one of Indian example and the other the challenge of fulfilling the needs of his Indian community.

7. We agree that the request of the Alberta Indian Association for at least one trained social worker on each reserve, is necessary to assist Indian families in making the many adjustments due to industrial impact and other stress. We disapprove of the present allocation of only one Social Worker for the whole province.

8. That the Alberta Government pursue agreement with the Dominion Government to the end that more provincial responsibility may be assumed for the education of Indian Children.

9. That we endorse the other six recommendations of the Cameron Report.

WOMEN OF UNIFARM
Forum Paper 1980

Women of Unifarm and its forerunners, Farm Women's Union of Alberta and United Farm Women of Alberta, have contributed a great deal to agriculture and to the betterment of farm families. Perhaps many have not realized the importance of Women of Unifarm as members of our Unifarm organization. Besides the many study papers – Laws for Albertans, Stress in the Farm Family, Coping with Stress and our program regarding the family, pension plans, etc., we have familiarized ourselves with laws and assisted Unifarm in carrying out its policies.

Perhaps we should spend a few minutes looking at the accomplishments of the past few years achieved by Women of Unifarm in co-operation with other organizations.

1. Farmers and small businessmen may deduct from income, salaries paid to spouses for income tax purposes.
2. Building of additional new hospitals and additions to existing ones.
3. New "small jurisdiction grant" of $750,000 for school districts with fewer than 1500 students.
4. $500,000 worth of pictures and facts about provincial animals and plants sent to Alberta schools.
5. Alberta Culture printed booklet "Women in Alberta".
6. Increase in support for school districts financially troubled by declining enrollments.
7. Incorporated family farms can be transferred to a son or daughter without payment of capital gains tax.
8. Legislation preventing non-Canadians from buying significant amounts of prime agricultural and recreational land.
9. A wife 60 years old can receive a supplement if her husband is over 65 and receiving Old Age Pension and supplement.
10. There is a provision to allow Canada Pension Plan contributors, who leave the labor force to raise young children, to drop out from the CPP benefit calculation, months of low or zero earnings incurred during the child-rearing period up to 7 years.
11. Adult education to include retraining for various trades for women as well as men.
12. Persons over 65 are eligible for Canada Pension whether they work or not.
13. Under Canada Pension, equality for spouses and dependent children of male and female contributors.
14. Removal of the $15 deductible for Blue Cross for senior citizens but 20% is payable by senior citizens to druggists for drugs.
15. Free Alberta Health Care for senior citizens.
16. Cost of living bonuses on social security.
17. Courses on co-operatives at the University of Alberta.
18. Privilege of cancellation of door-to-door contracts within 4 days.

We need a revival of locals so our grass-roots members can work for our organization and know what our problems are and work toward alleviating them.

The first farm organization was formed before Alberta became a province and early in the life of the United Farmers' of Alberta it was realized that there was a need for farm women, not only to have a social outlet from their isolation, but to play an important part in working toward betterment of farming communities. At first the farm women functioned as an auxiliary to the UFA but in 1916 there came into being a semi-autonomous organization – The United Farm Women of Alberta.

The women's organization dealt with matters related to health, education and the social betterment of rural areas. Standards of education, municipal hospitals, traveling clinics, mental health services, pensions and health services for senior citizens – all these things and many more have come about because of concentrated action of farm women. Protective legislation for women and children appeared on the statute books of Alberta earlier than in many other provinces due partly to the hard work and determination of farm women.

In 1951, recognizing the power of united world-wide efforts, the farm women became aa constituent society of Associated Country Women of the World.

In presenting annual submissions, sending letters to different levels of government, and taking care of issues that affect our rural areas, we in Women of Unifarm have done a great deal through the years, but we need to continue the work. Let us look at our goals and objectives and realize that we need our direct membership so that we will not let our forefathers down. The need for an active farm women's organization remains. There is work to do in a changing world and we must hold fast to the values established for us by women who led the way so many years ago.

Respectfully submitted,
Leda Jensen

HANDICRAFT LIST FOR THE F.W.U.A. CONVENTION 1961

Class 1: Sewing - 3 only
(a) Garment with smocking
(b) Practical housedress
(c) Plain apron
(d) Fancy apron
(e) Child's garment remade from used wool or similar material
(f) Girls best dress or ladies afternoon dress - any material
(g) Housecoat, kimona or duster coat
(h) Pyjamas (6 yrs and over)
(i) Sport shirt
(j) Jacket, blazer or coat - lined or half lined
(k) Ladies or girls winter skirt

Class 2: Knitting - 2 only
(a) Adults or childs cardigan) (light
(b) Adults or childs pullover) weight)
(c) Adults or childs heavy sweater
(d) Socks, solid color, plain or fancy stitch
(e) Socks, any pattern, 2 or more colors
(f) Shrug, shawl or stole
(g) Mitts, any age, any design
(h) Childs dress or suit - 2 to 5 yrs
(i) Infants knitted sets
(j) Knitted lace - edging, doily or any article (fine thread No 30-60)
(k) Knitted bed jacket

Class 3: Crochet - 2 only
(a) Dinner cloth 54" or more - all lace
(b) Lunch cloth - any size, crocheted edging or trim
(c) Crochet inserts on linen, not corners
(d) Crochet edge on 2 handkerchiefs
(e) Doily or tray cloth with crocheted edging or trim
(f) Crochet bed jacket

Class 4: Embroidery - 2 only
(a) Embroidery in white
(b) Embroidery in color
(c) Embroidery in cutwork
(d) Embroidery in cross stitch, not framed
(e) Drawn fabric embroidery or needle weaving

Class 5: Rug - 1 only
(a) Hooked rug - new material
(b) Hooked rug - salvage material
(c) Braided rug
(d) Any other variety - new material
(e) Any other variety - old material

Class 6: Loom Weaving - 1 only
(a) Small article e.g. bag, cushion cover, scarf or runner, etc.

(b) Large article e.g. bedspread, drapery, fabric length, rug, etc.

Class 7: Quilt - 1 only

(a) Quilt - patchwork, hand quilted

(b) Quilt - appliqued, hand quilted

(c) Cotton crib cover - hand quilted washable type padding

Class 8: Leathercraft & Copper work - 2 only

(a) Leather purse - not tooled or carved

(b) Tooled article

(c) Carved article

(d) Gloves

(e) Copper tooled article - framed picture, planter, lamp stand, tray, etc.

Class 9: Needlepoint & Pettipoint - 1 only

(a) Needlepoint - not mounted

(b) Pettipoint - not mounted

Class 10: Painting - 1 only

(a) Framed - painted in oils

(b) Framed - painted in water colors

(c) Framed picture embroidered, not needlepoint or pettipoint

Class 11: Other Handiwork - 1 only

(a) Cushion cover

(b) Stuffed toy

(c) Basketry - any article

(d) Machine embroidery - any article

(e) Machine quilting - any article, e.g. bedroom slippers, boudoir accessory

(f) Felt article

(g) Dried arrangement for table centre

Class 12: Bedspreads - 1 only

(a) Afghans - knitted or crocheted

(b) Bedspreads - knitted or crocheted

(c) Bedspreads - embroidered

(d) Bedspreads - appliqued

(e) Bedspreads - machine sewn

INSTRUCTIONS

1. Any local may exhibit 6 articles or less from any of the 12 classes listed.

2. All articles exhibited must be the work of our F.U.A., F.W.U.A. or junior members.

3. No prize winning exhibit from a previous F.W.U.A. exhibit may be entered.

4. One only means that only one article may be exhibited from that whole class.

5. Points for large articles such as rugs, quilts, table cloths, etc., will be 10 for first, 7 for second and 4 for third. Points for other articles 5, 3, & 1. A modified Danish system of judging will be used allowing the judge to award more than one article in a placing and/or award a higher or lower number of points according to workmanship and design of article.

6. No fancy work is to be mounted on colored paper, mounting may be sent for articles so that they may be mounted after they are judged.

7. No work is to be mounted unless specified, e.g. framed picture, in which case the entry is judged for the final effect of framing and frame.

8. Paintings on Commercial Number Charts disqualified.

9. Locals must request tags for exhibits from Central Office in October.

TEEN "B" SEMINAR AT GOLDEYE, July 22-28

I really enjoyed this year's seminar, working with a wonderful staff and group of young people. It was so much easier than the 1st time, probably because I knew what was expected of me.

The weather co-operated giving us 7 days of beautiful sunshine and giving us the opportunity to offer our whole slate of outdoor activities.

Having one bus arrive at 12:30 and the next one at 6:30 made the 1st day a long one.

There were 44 children, mostly from Calgary and a very good representation from the north, even one from High Level.

Our theme for the week was "Happy Days" and my group chose to be called "The Golden Bears", so I ended up with 9 cubs for the week.

I had three presentations to make.

1. Parliamentary Procedure
2. Duties of Officials
3. Reporting Back

I also had to look after the banquet committee.

In reporting back I tried to stress the reasons why Goldeye was built, and how important it was for them to show their appreciation by giving interesting reports on their return.

According to Jack Muza our group seemed to be the toughest one on demolishing ping-pong balls. It was very frustrating not to be able to supply the balls when they were locked up in the office. I would suggest that the co-ordinator be given the balls to give out or have the staff bring their own supply for the children.

Our staff of 7 with Keith Peppinck as co-ordinator, worked very well. We were fortunate to have Linda Schmidt (a former Goldeye student) who works for St. Johns Ambulance, to be our "Medic" for the week.

There were no serious mishaps and the children did get quite a bit of canoeing. Thanks to the beautiful weather I did not hear of anyone being bored.

It was a pleasure seeing Louise and Marlene at the May sessions.

Thank you for the opportunity to work on behalf of Women of Unifarm.

October, 1982

Respectfully submitted,
Olga Manderson

BULLETIN TOPICS

Bulletins covered a wide range of topics, for example:

1952 Bulletins:
Horticulture
Cooperation
Social Welfare
Legislation
Young people's work
Civil defense
Cultural activities

1961 Bulletins:
National health
Social welfare adoption in laws
Family Courts
Soil conservation
Community and property laws
Education

1963 Bulletins:
Women's rights
Education of gifted and handicapped children
Health Insurance
Co-op College
Agriculture and cooperation
Alberta Medical Health plan

Appendix C

Suggestive Program

Why follow a program?

The following program is submitted, not as a definite plan of work to be rigidly adhered to, but rather as a suggestive outline of activities for a Women's Section. The intention is that each Women's Section will get its board together, discuss the topics as presented, delete those not applicable to local needs and supplement those chosen with topics of vital interest to the community.

Time and experience have revealed that the Women's Sections that have made the greatest progress are those that have followed definite programmes of work. An organization to be successful must have a goal, a definite aim in view. This will necessitate its program committee thinking ahead, thinking not in a rut but broadly, comprehensively; considering the needs of those in the district and drafting plans to meet those needs; enlarging upon these plans until they include the physical, intellectual and moral uplift of the community, then mapping out a program that will cover the entire field.

In carrying out the year's work it should be remembered that the value of the program is not in the printed form, but in the thought behind the form and the force behind the thought. The best material may be engraved in gilded letters on manilla paper but without the force to translate that material into practical use it is worthless. Your Women's Section should be that force, the force that will put life and meaning into the printed program.

Program

Jan. 1st.—U.F.A. and U.F.W.A. New Year's Party. Original Stunts.

Jan. 15th.—Debate: "Resolved that Young People on the Farm Should Have an Allowance." Meeting conducted by Junior U.F.A. or Associate Members. Musical items.

Feb. 5th.—Joint Meeting with U.F.A. Reports of Delegates to Annual Convention. Community Singing. (Song sheets with words only may be secured from Central Office, price $1.25 per hundred; also community song book, containing words and music for 117 songs at $1.00 each.)

Feb. 19th.—Social gathering at a member's home. A survey of the community's needs. Round table talk on practical work to be taken up. Solo. Refreshments.

Mar. 5th.—Annual U.F.A. Banquet. Program arranged by school children. Community singing.

Mar. 19th.—Address by nearest doctor or nurse. Discussion:—"What can Our Local Do to Improve Public Health Conditions in the Community?"

April 2nd.—"My Plans for Flower and Vegetable Garden." Informal Discussion. Music. Refreshments.

April 16th.—Debate: "Resolved that there should be an Educational Qualification for the Franchise in Canada." Solo. Refreshments. (Debating material may be secured from the Department of Extension, University of Alberta, Edmonton South, on this subject.)

April 30th.—Address: "Training the Young People for Future Leaders." Discussion of the Junior Section of the U.F.A. Literature will be sent out from Central Office on request.

May 14th.—Public Meeting at School House. Arbor Day Program by Juniors. Tree Planting.

May 28th.—Paper or Address: Our Duty to the Foreign Immigrant." Community singing. Refreshments.

June 11th.—Address: "Laws Relating to Women and Children, with special attention to the Dower Act. (See "Legal Status of Women in Alberta. New edition will be on sale Jan. 1st, and may be obtained from Mrs. O. C. Edwards, Macleod.)

June 25th.—Annual U.F.A. Picnic. Special features for Junior Members.

July 9th.—Question Drawer. Plans for assisting the Rural Fair. Music. Reading. Refreshments.

July 23rd.—Rally with neighboring U.F.W.A. Locals. Discussion: "What Benefits do we Derive from the U.F.W.A.?" Special music. Community singing. Refreshments served out-of-doors.

Aug. 6th.—"Demontrations in Canning of Fruit and Vegetables." By one of the members.

Aug. 20th.—Graduation Exercises in Honor of Pupils Successful in Grade VIII. Examinations.

Sept. 3rd.—Address: "How can We Assist in the Educational Work of the Community." By the teacher or principal of the school. Plans made for Community Social at School House on Annual School Meeting secure full attendance.

Sept. 17th.—Rally of Junior Members. Address: "The Future of our Young People—How Are We Shaping It?" Plans for their year's activities. Musical numbers.

Oct. 8th.—Joint Meeting with U.F.A. Rally Night. Address: "History of Farmer's Movement in Canada; Aims and Future Possibilities." Discussion. Refreshments.

Oct. 22nd.—Address: "Instructions in Home Nursing and Lessons on Public Health." By Public Health Nurse. Or, Baby Clinic, babies to be judged by local doctor.

Nov. 5th.—Discussion—"Does the Course of Studies Meet the Needs of Rural Children?" Solo or other musical number. Discussion; "Is the present method of taxation for schools satisfactory?"

Nov. 19th.—Mock Parliament. Discussion on "How Tariff Affects Home-Making." Refreshments.

Dec. 10th.—Annual Meeting. Election of officers. Enrollment of members, payment of dues, appointment of convention delegates. Discussion on advisability of sending two or three young people in addition to duly-appointed delegates

Dec. 24.—Christmas Tree. Teachers and pupils of local schools the guests of the U.F.A. Program, gifts for children, refreshments, etc.

From *The U.F.A.*, Calgary, Alberta, April 15, 1922
By Mrs. R.B. Gunn, First Vice-President and Convenor of the
Educational Committee of the U.F.W.A.

An Editorial on Education

There is one phase of our work in which Alberta farm women evince perennial interest, and that is the subject of education. Perhaps this is because so many of our members flourished the pedagogical crayon before we succumbed to the lure of the western prairie and the importunities of its lonely bachelors.

Or it may be that in or quest for "whatsoever things are true," we turn with increasing interest to this subject, knowing that true education will ensure to our young people not only "preparation for a complete living," but will train them to live. The problems that meet boys and girls day by day are the ones of interest to them, not the ones they will meet in ten or fifteen years. And their ability to meet life's problems as they arise, in a spirit of unswerving truthfulness, courage, courtesy, justice and service will determine their value as citizens in later life.

Home, School and Community

Or finally it may be because we realize the close interdependence of home, school and community, and that all these factors act and react on the sensitive childish mind, moulding the destiny of the younger generation. In early primitive days boys and girls received the greater part of their education from the activities of the home and the social life of the neighborhood, and in slack occupational seasons imbibed learning by way of the three R's at the traditional little red schoolhouse. But with our present day development we have handed over to our teachers the training of our children not only in the three R's, but largely training also in the three H's—head, hand, and heart.

But even with our present day formal instruction the ideal arrangement is to have the activities of the home, school, and community so inter-related that it is hard to say where the influence of one begins and the other breaks off.

The Revised Course

In our study of school problems for a number of years we concentrated on the revision of the public school course. This revision has been under way for the past year, and the proposed course will no doubt be before the public and in the hands of teachers for next term. It is proposed to issue it as a tentative course for one year. And it will be the duty of all those interested in the education of our boys and girls to consider the proposed course in detail; to see that it is tried out in our schools fairly and conscientiously; to get information as to reasons for changes made or provisions retained in the new course which may not be in line with their own thought on the matter; and then submit to the Department of Education their criticisms or approval; so that, with a final revision, the new course may prove to be what the public has demanded. That is, a course suited to the needs of the young people of the Province for the next ten years; flexible enough to meet the varying demands of different sections of Alberta, one that will be the means of developing habits of accuracy, thoroughness, and speed in fundamental operations such as arithmetic, spelling, writing, etc., and one that will ensure the development of a strong, virile character, always bearing in mind that after all "it is not so much the kind of education a child receives but the spirit in which that education is used, and the purpose to which it is applied that counts."

A new phase of educational work came before our last Annual Convention, and that was the inspection of our rural schools. We passed a resolution on the subject which has paved the way for further discussion. For while individual members and Locals have casually discussed inspection for some time, this was the first definite plan brought before our Convention, with the result that many of our members are now actively discussing the resolution pro and con. And I think we should be prepared next Convention to reaffirm our position in the matter, or have amendments submitted to Central Office for distribution to all Locals, before December 1st. In this way our final stand will be a carefully considered expression of opinion from all sections of Alberta.

Causes of Retardation

The regrettable fact in connection with educational work is that, in spite of the interest manifested in the subject, so small a percentage of children finish Grade VIII. In rural schools this is usually due to retardation—children reach the age limit before completing the work of the first eight grades. It is estimated that in Saskatchewan 68.7 per cent. in the rural schools are from one to ten years behind their grade. In Foght's survey, "of 4,806 enrolled, 2,500 or more than half are in the first grade, and only 39 remain when Grade VIII. is reached." We have not yet received Alberta statistics, but we have reason to suppose that our report will be more than satisfactory.

One cause of retardation and non-attendance which accounts for a large percentage of case is that of ill health. Surely here is work for our organized farm women which they can carry out in their own communities. Several Locals have already instituted health clinics with marked success, others have had occasional medical inspection of schools. If the farm women would undertake a systematic study of physical education in relation to child health, and then with the very fine program of physical education outlined in the new curriculum for the guidance of teachers, if mothers and teachers would work together for the next ten years, we should see an army of young people more nearly approximating the ideal of "sane minds in sound bodies." And I know of no more potent force towards the attainment of this ideal than the teachers of this Province in co-operation with the organized farm women.

April 8th, 1929.

Mrs. Geo. G. Hobbs,
Secretary, Helmsdale Local 80,
Helmsdale, Alta.

Dear Mrs. Hobbs:

Congratulations on your new Local, to which has been given the number 80.

What a grand opportunity you women are going to have to work together for better rural conditions in your neighborhood. It will bring you into contact with forces, educational and otherwise which are tending to bind women closer and closer together for the purpose of improving those conditions under which we live. We cannot strike any very vital blow at these conditions singly, but "together", working in harmony with great numbers of other women, we can come closer to a higher social order.

Some of the broad lines of study in which the U.F.W.A. are engaged are covered by Health, Immigration, Legislation, Education, Social Service, Peace and Arbitration, and Co-operative Marketing. Each month a bulletin is sent out usually under one of these heads forming a basis for study in the Local.

We are forwarding you U.F.W.A. literature and shall be glad to have you call on us at any time for suggestions. Remember that Central is always interested. We are also sending you a receipt book and attach invoice for same. We are enclosing list of the supplies we keep at Central and shall be glad to forward any other supplies you require.

If any of the members for whom you remitted wish to receive the paper, please let us know and we shall be glad to add their names to the mailing list.

With all good wishes for the success of your Local.

Yours fraternally,
F. Bateman
Secretary.

BRIEF PRESENTED JUNE 7th, 1949 BY F.W.U.A.
EXECUTIVE TO THE ALBERTA HEALTH
SURVEY COMMITTEE

The F.W.U.A. Executive welcomes the opportunity of presenting to the Alberta Health Survey Committee some pertinent facts with reference to health. In this presentation we are concerned with health as applied to rural areas.

We wish to pay tribute to the National Department of Health and Welfare for its generous health grants to the provinces, and also, to the fine spirit of co-operation indicated by our Provincial Government. We recognize the present survey, and the allocation of health and hospital grants to be, in the words of the Minister of Health, Mr. Martin, as reported in Hansard April 5, 1949, "the first stage in a system of national medical and hospital care insurance." We stress the fact that, in our opinion, nothing less than National Health Insurance under the Dominion Government, in co-operation with the provinces, can provide adequate public health andmedical care service.

Our interest in health has two aspects, preventive and curative. We lay particular emphasis on the health of women and children, because of high maternal and infant mortality, and, also, because of the special health needs of children. Numerous recommendations dealt with subsequently in this report, concern an intensified program of prevention. Such a program could, over the years, materially reduce necessary capital expenditure on such institutions as mental hospitals, T.B. sanitoria, and to a lesser degree hospitals in general. In addition, such a preventive program could bring incalculable benefits to the general public by way of economic, social, and physical advantages resultant on a higher standard of health.

Rural Health

The situation with reference to health in rural areas is not a happy one, for example:

(a) Infant mortality in rural areas is higher than in adjacent cities.

(b) Maternal death rates are higher in rural areas than in adjoining urban centers.

The determining factor appears to be that health personnel tends to concentrate in urban centers with a consequent unequal distribution of doctors, nurses and dentists for rural areas.

Medical Inspection for Rural Children Entering School

Health records of doctors and nurses in the city show that children arrive at school age, (We quote from *Canada's Health*, Auspices National Committee for Mental Hygiene) "With a multitude of disabilities, defective eyesight, diseased tonsils, rotten teeth, curved spines, rickets, etc." But no general provision for a health check-up is made for rural children entering school for the first time. We feel that a complete medical check-up, with corrective treatment where necessary, should be a minimum health objective.

Dental caries

We draw attention to an article in *The Woman's Home Companion*, July, 1948, entitled, "The Simplest Health Program Ever Sponsored". This proposal concerns the elimination of tooth decay by technicians equipped to swab children's teeth periodically with sodium fluoride. Dr. McLaren, of the Department of National Health and Welfare, Ottawa, writing in

Health Welfare, Vol. 4, No. 4, January 1949, says that this treatment ushers in "A New Era in Dental Health".

T.B.

We are alarmed at the incidence of T.B. both among children and adults. The travelling T.B. units are doing excellent work. We need more of them, and the province should be covered at regular intervals, in order to insure early detection.

Bovine tuberculosis, or T.B. arising from bovine infection, represents a significant proportion of all T.B. cases. T.B. tested herds are relatively rare. This is especially true in the case of small farm herds. Thus, the use of pasteurized milk can not be too greatly stressed. In addition, unpasteurized milk may be a spreader of undulant fever, septic throat, and other diseases. It is suspected also, that it may be the source of many vague ills, discomforts, and under par health conditions.

Contagious Diseases

Whooping cough according to *Hygeia* is actually the most serious of childhood disorders: to quote, "where it does not kill it may wake the drowsing tiger of T.B., be the forerunner of asthma, or so injure the sympathetic and endocrine systems that a child grows up lacking in energy", etc. The article concludes: "Whooping cough is a preventable disease. Thanks to brilliant work done in recent years an effective vaccine has been developed that actually prevents this terrible disease". The same thing is relatively true of scarlet fever and diptheria. Thus, we feel that we should have an immunization program which assures all children immunity against small pox, diptheria, scarlet fever, and whooping cough.

Mental Hygiene and Child Guidance

We are strongly of the opinion that there is a great need for mental hygiene and child guidance clinics for rural areas. The doctors and nurses who act as health inspectors of schools should have special training in psychiatry in order to recognize the onset of mental ills. It is now recognized that many cases of mental ill health have their roots in childhood maladjustments, and that fears, unhappiness and anti-social behavior may be expressions of mental ill health.

We realize that many factors contribute to the health and well being of a family.

1. A healthy environment - sanitary and spacious with provision for relaxation. This aspect of health, we assume, does not come within the purview of your committee, except insofar as health education may contribute to this end.
2. Nutrition - concerns the promotion of growth and repair of waste in organic bodies, while hunger may be satisfied by ample food. Yet a 'hidden hunger' may exist and result in grave malnutrition in the midst of food abundance. Someone has said that "the kitchen sink is the best fed mouth in America." Definitely there is need for education with reference to food values.
3. Health guidance - is a large order and covers the field of child care and a study of emotional growth and development as a base, which broadens to include the family as a unit, and the community as a whole.
4. Prenatal and post natal care - are of primary importance in order to reduce the tragic toll of morbidity and mortality. We think this important phase of public health requires no elaboration at this late date.

In conclusion we summarize the health needs of this province:

We need:

(a) Provincial wide coverage by health units to carry out the following health services:
1. Vital statistics and records.
2. Communicable disease control.
3. T.B. control.
4. Mental hygiene and child guidance clinics.
5. Prenatal and post natal services.
6. Medical health inspection of school children, and necessary corrective treatment to implement its findings.
7. General health education.

We need:

(b) Municipal doctors to provide:
1. Medical care.
2. Obstetrical care.
3. Minor surgery.
4. To act as medical health officers.
5. To organize immunization centres.
6. To inspect schools and examine school children.
7. To provide prenatal and post natal care.
Arising out of (a) and (b)

We need:

(c) More doctors, more nurses, more trained personnel.

To encourage suitable young people to take up these professions we suggest, (in addition to professional training grants) governmental assistance by way of scholarships or bursaries.

All of which is respectfully submitted.

> Vera Lowe, President
> Mary B. Pharis, 1st Vice-President
> Susan M. Gunn, 2nd Vice-President

To the Premier of Alberta and Members of the Executive Council, January 19th, 1959.

Mr. Chairman and Members of the Council:-

The Farm Women's Union of Alberta appreciate this opportunity to meet with you to present problems which have been brought to our attention by farm women from all parts of the province meeting in annual convention. We wish to commend our Government for the Five Year Plan announced last year which provides Homes for the Aged, etc. The new Institution for Delinquent Girls and the fine courses provided for rehabilitation. For the educational opportunities to the many groups in our society. We look forward to the report of the Cameron Commission and the implementation of the recommendations of the Blackstock Commission.

We will try to present our viewpoints briefly and clearly, and will welcome your comments.

1. Property Laws Affecting Women

(a) For many years the Farm Women's Union of Alberta have been requesting the Provincial Government to introduce Community Property Laws. We therefore would like to know Mr. Premier if you and your Government have considered the introduction of such laws. If not, we would like to recommend to you, Sir, that this cabinet appoint a committee to study all aspects of Community Property Laws. This committee, we suggest, should investigate fully; then publish their findings referring to the feasibility of establishing Community Property Laws.

(b) And further we request that the Intestate Succession Act be amended as to Section 3 to provide that where an intestate dies leaving a widow and children over 21 years of age, the widow shall be entitled to the estate or share thereof up to $20,000.00 before any distribution is made to the children, and that the children's share as set out in Section 3 shall only apply to the amount of an estate over $20,000.00 in value.

2. Equality for Women

We the Farm Women of Alberta respectfully ask the Provincial Government to appoint women magistrates wherever feasible and to use women jurists in civic courts.

3. Health

The rural Health Units have benefitted many rural people. We would ask our Government to see that all rural Health Units are fully staffed, and that Mental Health Services be added to all rural full time Health Units.

4. Provincial Health Scheme

We commend the Government for setting up the Provincial Health Scheme and for their co-operation towards a National Health Scheme.

5. Increased Staff for Welfare Department

We wish to commend our Government for the installation of the fourteen welfare officers and their assistants throughout Alberta and we strongly urge that you proceed with the program in areas which are not sufficiently serviced at the present time.

6. Education

From resolutions presented to our annual conventions the F.W.U.A. continue to suggest that the total cost of education be the responsibility of major governments, less cost of erection and maintenance of buildings; and that where possible basic material in text books be standardized in all provinces.

7. Semester System

The present three semester system at the Red Deer Composite High School has proved to be very valuable to many students and we would urge its continuation.

8. Compulsory Refresher Course

We feel that many older teachers find it difficult to change their teaching methods, and new methods are required to keep up with the changes in our society. We would request the Provincial Government to have all teachers take a refresher course at least once in every ten years.

9. Heaters in School Buses

Many parents in sparsely settled areas have expressed grave concern over the possibility of engine failures on school buses which would cause heaters to quit operating. Therefore we would request that a secondary heating unit be installed for emergencies in these buses that travel in sparsely populated areas.

10. Examinations for Grades X and XI

We believe that students in Grades X and XI should write final examinations, therefore we would like to suggest that standardized basic examinations for these grades be prepared by the Department of Education in the basic subjects and be corrected by teachers, other than their own, within the School Division.

11. Departmental Examinations for Mathematics 31

Since Mathematics are required for a matriculation subject by some University Faculties we request that departmental examinations be given in Mathematics 31.

12. Credits for 4-H Work

In view of the excellent program and work being done in the 4-H groups we would ask the Department of Agriculture to work out with the Department of Education a system of credits to be granted for 4-H members for the work they do during the year, such as credits are given for extra curricular music.

13. Agricultural Economics and Farm Management

We request the Board of Governors of the University of Alberta to establish a school of Agricultural Economics and Farm Management within the Faculty of Agriculture.

14. Schools of Agriculture

In order to determine the present and future requirements of our Schools of Agriculture, we would ask our Provincial Government to make an immediate survey of all departments of the Schools of Agriculture and Home Economics, and without delay to carry out necessary construction, renovation and purchase of equipment to meet and maintain a standard in keeping with the importance of agriculture.

15. Protection of Farm Lands

Farmers who live adjacent to large centres are having their private property rights violated by persons entering and causing willful distruction of farm property, machinery and animals. The present laws are either inadequate or not properly enforced to protect owners. We bring this serious situation to the attention of our Government, with the hope that something will be done very soon to correct this unfair treatment of our farmers.

16. Sale of Fireworks

The serious injuries and property damage cause by careless use of fireworks has been brought to our attention and therefore the F.W.U.A. would request our Provincial Government to prohibit the sale of fire works by law, excepting only public displays conducted by qualified persons.

Conclusion

Once again we thank you for your comments and interest in our brief. We appreciate this opportunity of meeting with you each year to discuss the problems brought before our convention. We thank you for your consideration of our requests in other years and hope that you will give consideration to this brief.

Respectfully submitted,
The Farm Women's Union of Alberta.

Memorandum to Hon. Guy Favreau, Minister of Justice
March 3, 1965

The Farm Women's Union of Alberta has since 1946 requested that Canadian Divorce Laws be reformed. For a period of years our policy was basically the British Divorce Laws of 1937 re-worded slightly at one time to conform to the request of the Canadian Bar Association. Three years ago a committee of the provincial F.W.U.A. Board made a thorough study of divorce laws in various countries, following which the annual convention accepted the following policy:-

Divorce

That the Divorce Laws be revised so that divorces may be granted for the following reasons:-

(a) Habitual drunkeness or habitual intoxication by reason of taking or using to excess narcotics or stimulating drugs or preparations for a period of not less than two years or has been a habitual drunkard or habitually been so intoxicated for a part or parts of such period.

(b) Adultery.

(c) Desertion without cause for a period of three years immediately preceding the Petition.

(d) If since the celebration of marriage one spouse has treated the other party with cruelty.

(e) If one party is incurably of unsound mind and has been under treatment for five years immediately preceding the Petition.

(f) The wife may petition on grounds that the husband has been guilty of rape, sodomy or bestiality.

(g) Legal presumption of death of the other spouse.

Throughout the years the Farmers' Union of Alberta have also supported this policy.

The Women's Conference of the Manitoba Farmers Union also passed this resolution adding

"WHEREAS the divorce laws of Canada are out-dated (dating back to 1857 with few changes),

THEREFORE BE IT RESOLVED that we petition the Government of Canada to establish one Canadian domicile rather than provincial domiciles, and accept foreign marriage certificates at face value."

This was supported by the Manitoba Farmers Union annual convention.

While the F.W.U.A. did not include this clause in their resolution they were aware of the implications and difficulties created by the necessity of establishing the domicile of the husband.

We petition the Canadian Government to reform the Canadian Divorce Laws.

Respectfully submitted by
Mrs. Russell Johnson, President,
FARM WOMEN'S UNION OF ALBERTA

SUBMISSION TO THE PRIME MINISTER OF CANADA AND MEMBERS OF THE CABINET ON THE REPORT OF THE ROYAL COMMISSION ON THE STATUS OF WOMEN BY WOMEN OF UNIFARM
March 10, 1972

Mr. Prime Minister and Members of the Cabinet:

We, the Women of Unifarm, appreciate the opportunity of presenting for your consideration the views of our organization on many topics of pertinent interest to all sectors of the Canadian society today. We are a province wide organization and represent a rural viewpoint since we are the wives of active farmers or ranchers, or active farmers or ranchers in our own right.

In October 1972 the Women of Unifarm instituted a study and educational program for all rural residents of the province of Alberta on the Report of the Royal Commission on the Status of Women. Initially we held twenty eight meetings at widely scattered points in our province, and six hundred and seventy three key people participated in these meetings. They returned to their own communities and many groups were stimulated to set up study programs. An on-going program evolved, and our speakers are still very much in demand for continuing study and discussion. We have evaluated our meetings, and the following represents the views of the majority of our membership.

We hope that your government will evince a continuing interest in the Report of the Royal Commission on the Status of Women, and will prepare and present to the House the necessary legislation in accordance with the wishes of the majority of Canadians.

Equality of Opportunity

Although the laws of our country say that there shall be no discrimination in regard to sex in employment, and that there shall be equal pay for equal work, these laws seem to be very difficult to enforce. Nevertheless there are loopholes in the law, which must be reviewed and revised, so that no employer can use discriminatory practices. Part of the difficulty of enforcement must be laid at the door of women themselves, many of whom are afraid to complain lest there be ill will and retributions. Firmly entrenched attitudes of sex typing are a contributory factor to discrimination, and there, our government itself has been one of the prime offenders. The government must review its own attitudes towards discriminatory practices. Equality of job opportunities and promotion to senior positions must be available to qualified women of intelligence and capability. With opportunity and educational training women can make a very significant contribution to the economic life of our society.

In particular educational and re-training opportunities are even more restricted for rural women than they are for urban residents. We strongly feel that rural women must be given the same opportunity as men to participate in re-training and rehabilitation programs such as those sponsored by A.R.D.A. and Manpower. The many small farm units that are presently in very serious economic difficulties could be assisted into viability in many cases, with the re-training and employment of the farm wife.

The Family

We believe that the ideals and behaviour of our society must be influenced through the family. There can be no question that the history of moral decadence of other societies has been caused by the decay of family living. The concept of a strong Canadian family must be supported by positive action on the part of our governments.

We strongly support that broad programs of family life education, including sex education be taught in the elementary and secondary schools, that these courses be taught by professional people fully qualified to do so, and that boys and girls be taught in the same classroom.

The sole support parent, including unmarried mothers, must receive assistance in counselling so that they may overcome social and emotional problems, and so fill their dual role as capably as possible. We recommend a guaranteed annual income for sole support parents. For the sole support parent we fully support the concept of day care centres, and that the fees for these children be fixed on a sliding scale based on the means of the parent.

We strongly support the concept that the guaranteed income supplement to Old Age Security benefits be increased so that the annual income of the recipients is maintained above poverty level.

Full time housekeepers and visiting homemakers are few in numbers in Canada. We recommend that training programs for such workers be expanded, so that these women are competent and responsible. In cases of a large family a housekeeper may be maintained in a home at less cost than putting the children in a day care centre. If the means of the parent cannot supply an adequate housekeeper, we believe that there must be some government assistance, again on a sliding scale. We do not support the concept that housekeepers and day care centres should be made available to all working parents - only in case of proven need.

Homemaker services, Victorian Order of Nurses, Meals on Wheels and other visiting services very often make it possible for elderly people to continue to live in their own homes, and we wish to see these services continued and expanded. These services fulfil a dual function - the senior citizens are usually much happier in their own familiar surroundings, and this relieves the congestion in nursing homes and auxiliary hospitals.

We support the premise that a wife who is financially able to do so may be held to support her husband and children in the same way that the husband may now be held to support his wife and children.

Married women are placed in unsatisfactory and humiliating positions in the granting of credit - in that a husband must sign a contract made by his wife for credit or loan, regardless of the security a wife may hold in her own name. In most cases credit cards are issued only in the name of the husband. We protest that, on marriage, a woman assumes her husband's credit rating, regardless of the credit rating she may have established in her own name. Sex should not be a factor in reaching credit and loan decisions, and laws should so clearly state.

We are concerned about the plight of native women in our society. We would ask that the Indian Act be revised, so that native women who marry non-natives shall have treaty rights restored for themselves and their children, and that this be retroactive. We ask that there be continued and expanded support for halfway houses for native women, who are the victims of our society, because of alcohol, drugs and prostitution.

While this has no direct relation to the Status Report, other than that family life suffers emotional stress and financial hardship when the bread winner is involved in labor disputes, we would ask that a review be made of Labor laws and negotiations. Our national economy has suffered repeated and heavy blows caused by strikes. We would ask that in the case of essential services (police, postal, transit, dock employees, firemen, etc.) that laws be revised so that services must continue while arbitration proceeds.

Family Planning

It is natural for a man and a woman to want children, giving them affection, care and training. Every child has the right to be wanted and cherished, and when a child is unwanted and unloved there is tragedy for that child. There are too many unwanted and unloved children in our society.

It is absolutely essential that people must be educated in family planning, and that education must be especially geared to the needs of the people who most need it. In addition to family planning education there must be counselling and the free provision of contraceptives in cases of need. These contraceptives, including intra uterine devices, diaphragms and birth control pills should be available when prescribed by a physician, and should be available to minors without parental consent.

We support the use of sterilization as a method of birth control, especially so when there are enough children in the family so that there would be difficulty in providing food, clothing, shelter and education for more. Vasectomy or tubal ligation must be the choice of the parents involved.

Abortion must not be used as a method of birth control, and we are extremely opposed to "abortion on demand". Repeated abortions have physical and psychological effects on a woman, and very often have a damaging guilt complex that can be very destructive to family relationships.

The question of abortion should be settled between a woman and her physician, and we favor the amendment of the Criminal code so as to permit a qualified practitioner to perform an abortion if the woman has been pregnant for twelve weeks or less. The long delays caused by application to an abortion committee can exceed the twelve week period, and we recommend that hospital abortion committees be abolished.

Rural women are especially discriminated against because in many areas there are no hospitals, accredited, approved or otherwise, sometimes there is not even a qualified practitioner within reasonable distance. We support the idea of mobile clinics, so that rural families may receive medical advice and assistance when it is needed.

Divorce

We recognize and strongly support the breakdown of marriage as grounds for divorce, but we do not support the three year separation period, nor the five year period by the petitioner's desertion of the respondent. Common law relationships and illegitimate children too often are the result of this prolonged waiting period, and we strongly support the concept of a one year separation period.

Legal fees for divorce are still too high, and completely outside the reach of women on social assistance. Separated and deserted wives who are in financial need must be provided with assistance to obtain divorce. We support the recommendation for Family Courts in those provinces that do not have them, and that these courts shall have jurisdiction over the assessment and payment of alimony and maintenance.

In the case of divorce the children are nearly always placed in the custody of the mother. This seems to assume that Motherhood is sacred - yet many women are not good mothers. Very often the father is a more responsible and capable person, and relegating the care of the children to the mother is discriminatory to men. Although we realize that custody is a difficult matter to legislate, the judges of our courts must be instructed to give more serious consideration to the welfare of the children involved when deciding custody.

We support very strongly that the Divorce Act be amended so that parents are responsible for the maintenance of their children over the age of 16 as long as they are in secondary school.

Farm women are, in most cases, a true working partner on the farm, contributing labor in house, field and barnyard as well as secretarial and bookkeeping services. It is grossly unfair that upon the dissolution of a marriage, there is no provision made for the financial security of the woman who has made such a contribution, except by the good will of the husband. We strongly support that upon dissolution of a marriage each partner shall have a right to an equal share in the assets accumulated during marriage.

Canada Pension Plan

Canada Pension Plan discriminates against women because only those women who work outside the home in paid employment may claim benefits. We support the concept that a housewife, who is also a producer of goods and services, and makes her contribution to society as a homemaker, should be allowed to contribute as a self-employed worker, and the amount of her contributions should be one of self choice.

If a homemaker chooses not to contribute in this manner, and becomes widowed, her survivor benefits should continue even if she re-marries. In the event her second husband pre-deceases her, she should have the right to claim only one pension, but it should be that of her first or second husband as she so chooses.

A woman who works outside the home on a part time basis, and who therefore contributes and amount related to earnings, should have the right, if she so desires, to increase her contributions to obtain eligibility for the maximum pension benefit possible if she was a full time worker.

Due to the fact that it is becoming increasingly difficult to obtain competent, reliable farm labor, many rural women accept a double work load (that of a homemaker and a farm laborer). These women cannot claim remuneration for work performed in fields and farm yard, nor, if the husband is willing to pay for such services, can he deduct such wages from his income tax as a legitimate farm expense. This is rank discrimination against rural families, and we, as rural women, ask that this situation be legally rectified and the employment of one spouse by another should be regarded as payable, pensionable and deductable employment.

Women in Public Life

We deplore the present lack of involvement of women in all aspects of public life. The extreme imbalance in government committees, boards, commissions, the provincial and federal Houses, the Senate and the judiciary indicates a social pattern that results in eliminating women from important offices.

We strongly support the recommendation that two qualified women from each province be appointed to the Senate as seats become vacant, and that women continue to be summoned until a balance is achieved. And we further desire that not only professional urban women be included in the Senate, but that consideration be given to rural women who are fully capable of presenting the views of the agricultural segment of Canadian society. Because women, very often, are not financially independent of their husbands, the financial qualifications for Senate memberships should be abolished.

The Provincial and Federal governments must realize that competent women can make efficient and effective contributions to the judiciary, government boards and commissions, and that women, representing half the voting population of Canada, must be given the opportunity to make that contribution.

The adoption of reform proposals to ease the financial burden for all candidates who compete in elections would undoubtedly bring forth more female political aspirants.

We support the concept that in all cases women should carry the same responsibility to perform jury duty as men do.

Women in Voluntary Organizations

Women contribute many million service hours annually in voluntary organizations, which offer services of many kinds to the general public. While it is desirable that citizens themselves assume some responsibility for meeting the needs of society, many voluntary organizations find themselves in serious financial difficulties.

We believe that grants should be provided to associations performing useful functions of particular concern to women and the family - particularly educational and upgrading services and projects of public interest.

We believe also that volunteer work should be recognized as experience for employment purposes, provided that the volunteer experience is relevant to the requirements of the paid position.

In conclusion, we do not ask for special privileges for women - only that discriminatory practices cease, and that we be accepted as persons of equal intelligence and abilities, and be given equal opportunities.

Respectfully submitted,
Women of Unifarm
March 10, 1972

SUBMISSION TO THE SPECIAL COMMITTEE ON CHILD CARE
June 1986

MEMBERS OF THE TASK FORCE

We are pleased to have been asked to participate in the hearings of the Special Committee on Child Care.

The Women of Unifarm is a provincial organization composed of farm women. We have a structure in place that enabled us to conduct an opinion poll on a province wide basis to assist us in reaching our conclusion.

The majority of women contacted were between 20 and 40 years of age, and the majority of their children were under 10 years of age. We attempted to ascertain child care needs, what was available in their area and what they believe is required in their community.

REASONS FOR NEEDING DAY CARE

Eighty-eight percent of those polled said that day care should be available for the following reasons:

1. Child care is needed at peak time of seasonal work i.e. haying, seeding, harvesting and other on-farm work.
2. Full-time day care for those whose parents work off the farm and after school care for young children.
3. When business or medical appointments take the parent(s) away from home.
4. Occasional care for children with special needs.

WHAT IS CURRENTLY AVAILABLE

1. In some areas there aren't any child care services available.
2. In other areas many rely on co-operative arrangements with neighbors or relatives who provide child care services. (However, obviously during the busy seasons on the farm, most mothers are busy at the same time.)
3. Some urban centres have small day care services in private homes but many of them are too far away from the farm to be practical for part-time care.

WHAT DAY CARE SERVICES ARE REQUIRED?

Day care needs in rural areas differ from urban areas because of distances, sparse populations and the need for child care during peak seasonal work. Although parents employed full time off the farm have access to day care in larger centres there is still need for after school care for young school children arriving home on school buses.

We suggest that rural child care needs can be filled as follows:

1. Encourage the development of satelite day care centres in private homes with limited enrollment for full time child care of about 3 to 4 children.
2. Community Child Care Co-operatives made up of participating parents and operated on an exchange of services.
3. Develop a Central Registry of available and responsible persons who are prepared to provide child care on a part time basis in the child's home or in their own homes.

4. Day care services could be offered in existing facilities such as community halls or unused classrooms situated within the farming community.
5. Day care workers should be encouraged to take some training in first aid and understanding child development.

Any of these services could be used to provide child care for young children from birth upward. As well, they could be used to provide after school care for the youngsters travelling on school buses. By pre-arrangement the bus could stop at the child care facility.

It is interesting to note that of those polled five times as many requested seasonal care for pre-school children and after school care than did those requesting all year care.

SUMMARY

The problem of rural day care is receiving more attention because of the economic decline in rural areas and the devastating effect this is having on rural communities and family farms. Many rural mothers are finding they have no alternative but to work on the farm or to seek off-farm employment to supplement the farm income. Whichever way she does it - the young children are a concern - often left unattended through no choice of the parents and in grave danger from the equipment and livestock on a farm.

An affordable and flexible day care system has become a must for rural communities.

At this time it may be inappropriate to suggest methods of financing any of our suggestions as some of them are based on self-help. However, depending upon which method might be chosen, we would be prepared to offer further input.

Respectfully submitted,
WOMEN OF UNIFARM

Index